Springer Proceedings in Mathematics & Statistics

Volume 516

This book series features volumes composed of selected contributions from workshops and conferences in all areas of current research in mathematics and statistics, including data science, operations research and optimization. In addition to an overall evaluation of the interest, scientific quality, and timeliness of each proposal at the hands of the publisher, individual contributions are all refereed to the high quality standards of leading journals in the field. Thus, this series provides the research community with well-edited, authoritative reports on developments in the most exciting areas of mathematical and statistical research today.

Amin Bahmanian
Editor

Combinatorics and Combinatorial Computing

MCCCC34, Normal, IL, USA, October 21–23, 2022

Springer

Editor
Amin Bahmanian
Department of Mathematics
Illinois State University
Normal, IL, USA

Midwestern Conference on Combinatorics and Combinatorial Computing mcccc mcccc2022 MCCCC 34 Normal, IL, USA 2022 10 21 2022 10 23 https://about.illinoisstate.edu/mcccc34/

ISSN 2194-1009 ISSN 2194-1017 (electronic)
Springer Proceedings in Mathematics & Statistics
ISBN 978-3-032-08548-1 ISBN 978-3-032-08549-8 (eBook)
https://doi.org/10.1007/978-3-032-08549-8

Mathematics Subject Classification: 05B10, 05C15, 13A05, 68R10, 05-06, 11P99, 05C78, 05B10, 05A15, 05A19, 05C65, 05C25, 05B40, 05C70, 13F15

This Springer imprint is published by the registered company Springer Nature Switzerland AG
The registered company address is: Gewerbestrasse 11, 6330 Cham, Switzerland

Preface

The Midwestern Conference on Combinatorics and Combinatorial Computing (MCCCC) originally started as the Carbondale Combinatorics Conferences and were held (1986–1990) at the Southern Illinois University, Carbondale. The conference has since evolved to a prestigious national conference in the areas of Combinatorics and Computing and has achieved recognition and attracted scholars and researchers of various backgrounds from different institutions. The 34th Midwestern Conference on Combinatorics and Combinatorial Computing took place on October 21–23, 2022, at Illinois State University, Normal, IL, USA. The Program Committee was formed by Guantao Chen (Georgia State University, USA), Saad El-Zanati (Illinois State University, USA), Heather Jordon (Mathematical Reviews, USA), Gyula O. H. Katona (Alfréd Rényi Institute of Mathematics of the Hungarian Academy of Sciences, Hungary), Barbara Maenhaut (University of Queensland, Australia), and David Pike (Memorial University of Newfoundland, Canada). The conference organizers were Amin Bahmanian (Illinois State University, USA), Ryan Bunge (Illinois State University, USA), Craig Cullen (Illinois State University, USA), Heather Jordon (Mathematical Reviews, USA), Mike Plantholt (Illinois State University, USA), Dan Roberts (Illinois Wesleyan University, USA), Songling Shan (Illinois State University, USA), and Papa Sissokho (Illinois State University, USA).

The conference brought together a total of 97 participants, including a number of high school students, mathematics teachers, undergraduate students, and PhD students. The list of invited speakers consisted of the following:

- Brian Alspach, University of Newcastle, Australia
- Marijn Heule, Carnegie Mellon University, USA
- Mariusz Meszka, AGH University of Science and Technology, Poland
- Anita Pasotti, University of Brescia, Italy
- Wannasiri Wannasit, Chiang Mai University, Thailand
- Anna Weigandt, Massachusetts Institute of Technology, USA

The conference coincided with a celebration of Saad El-Zanati who retired in August 2021 after teaching for 30 years at Illinois State University. Saad earned his PhD from Auburn University in 1991, and in the same year he began as an assistant

Alaittin Kirtisoglu, Illinois Institute of Technology, USA
Donald L. Kreher, Michigan Technological University, USA
Yi-Lin Lee, Indiana University at Bloomington, USA
Yifei Li, University of Illinois at Springfield, USA
Alison Marr, Southwestern University, USA
Dan McQuillan, Norwich University, USA
James McQuillan, Western Illinois University, USA
John McSorley, Southern Illinois University Carbondale, USA
Mariusz Meszka, AGH University of Science and Technology, Poland
Trent Minder, Southern Illinois University Edwardsville, USA
Mojtaba Moniri, Normandale Community College, USA
Jeffrey Mudrock, College of Lake County, USA
David E. Narvaez, University of Rochester, USA
Evan Natsis, El Paso Gridley High School, USA
Levi Neiburger, Normal Community High School, USA
Nicholas Newman, Troy University, USA
Tung T. Nguyen, Western University/Onepick Inc, USA
Matt Noble, Middle Georgia State University, USA
Adam O'Neal, University of Kentucky, USA
Anita Pasotti, University of Brescia, Italy
Jawahar Pathak, Lincoln University, USA
Katherine Perry, Soka University of America, USA
Mike Plantholt, Illinois State University, USA
Zahid Raza, University of Sharjah, UAE
Anja Remshagen, University of West Georgia, USA
Williem Rizer, University of Kentucky, USA
Dan Roberts, Illinois Wesleyan University, USA
Chris Rodger, Auburn University, USA
Michael Schroeder, Marshall University, USA
Andrew Schwartz, Southeast Missouri State University, USA
Michael Severino, Illinois State University, USA
Hossein Shahmohamad, Rochester Institute of Technology, USA
Songling Shan, Illinois State University, USA
Gunjan Sharma, Illinois Institute of Technology, USA
Rinovia Simanjuntak, Institut Teknologi Bandung, Indonesia
Smriti Singh, DePaul University, USA
Papa Sissokho, Illinois State University, USA
Kayla Smock, Bloomington High School, USA
Rachel Taylor, DePaul University, USA
Tommaso Traetta, University of Brescia, Italy
Kevin Treat, U.S. Air Force Academy, USA
Jenna Varnell, North Carolina State University, USA
Wannasiri Wannasit, Chiang Mai University, Thailand
Anna Weigandt, Massachusetts Institute of Technology, USA
Larry Wilson, Center for Communications Research, La Jolla, USA

Nicholas Winschel, University of Virginia, USA
Bryant Xie, University of Arkansas, USA
Guoning Yu, Georgia State University, USA
Bailee Zacovic, University of Notre Dame, USA
Stephanie Zeppetello, East Leyden High School, USA

All papers in this volume were refereed. We are deeply grateful to all our referees for their time-consuming effort and discipline in evaluating the articles. We thank Remi Lodh from Springer for his enthusiasm and for his valuable assistance. We would like to recognize the US National Science Foundation, the Institute of Combinatorics and its Applications, and Illinois State University (the Department of Mathematics, the College of Arts and Sciences, the Office of Research and Graduate Studies) for their financial support. We would like to thank Illinois State University for hosting this conference. We also address our gratitude to the fellow local organizers Ryan Bunge, Craig Cullen, Heather Jordon, Mike Plantholt, Dan Roberts, Songling Shan, and Papa Sissokho, and the Program Committee Guantao Chen (Georgia State University, USA), Saad El-Zanati (Illinois State University, USA), Heather Jordon (Mathematical Reviews, USA), Gyula O. H. Katona (Alfréd Rényi Institute of Mathematics of the Hungarian Academy of Sciences, Hungary), Barbara Maenhaut (University of Queensland, Australia), and David Pike (Memorial University of Newfoundland, Canada).

Normal, IL, USA Amin Bahmanian

Contents

Astrid Lilly Reed College, Portland, OR, USA

John P. McSorley Southern Illinois University, Carbondale, IL, USA

Steven J. Miller Williams College, Williamstown, MA, USA

Tracer Mills Illinois State University, Normal, IL, USA

Anthony Netz Oregon State University, Corvallis, OR, USA

Prakod Ngamlamai Harvey Mudd College, Claremont, CA, USA

Lauren Nowak Leadership High School, San Francisco, CA, USA

M. A. Ollis Emerson College, Boston, MA, USA

Jawahar Pathak Lincoln University, Oakland, CA, USA

Matthew Phang Williams College, Williamstown, MA, USA

Dan Roberts Illinois Wesleyan University, Bloomington, IL, USA

Chris Rodger Auburn University, Auburn, AL, USA

Kaysie Scott University of Minnesota Duluth, Duluth, MN, USA

Patrick Ward Illinois Wesleyan University, Bloomington, IL, USA

Larry Wilson Center for Communications Research – La Jolla, San Diego, CA, USA

Sum and Difference Sets in Generalized Dihedral Groups

Ruben Ascoli, Justin Cheigh, Guilherme Zeus Dantas e Moura, Ryan Jeong, Andrew Keisling, Astrid Lilly, Steven J. Miller, Prakod Ngamlamai, and Matthew Phang

Abstract Given a group G, we say that a set $A \subseteq G$ has more sums than differences (MSTD) if $|A+A| > |A-A|$, has more differences than sums (MDTS) if $|A+A| < |A-A|$, or is sum-difference balanced if $|A+A| = |A-A|$. A problem of recent interest has been to understand the frequencies of these type of subsets. The seventh author and Vissuet studied the problem for arbitrary finite groups G and proved that almost all subsets $A \subseteq G$ are sum-difference balanced as $|G| \to \infty$. For the dihedral group D_{2n}, they conjectured that of the remaining sets, most are MSTD, i.e., there are more MSTD sets than MDTS sets. Some progress on this conjecture was made by Haviland et al. in 2020, when they introduced the idea of partitioning the subsets by size: if, for each m, there are more MSTD subsets of D_{2n} of size m than MDTS subsets of size m, then the conjecture follows. We extend the conjecture

R. Ascoli (✉)
Princeton University, Princeton, NJ, USA
e-mail: rascoli@princeton.edu

J. Cheigh · S. J. Miller · M. Phang
Williams College, Williamstown, MA, USA
e-mail: jhc5@williams.edu; sjm1@williams.edu; mlp6@williams.edu

G. Z. Dantas e Moura
Haverford College, Haverford, PA, USA

R. Jeong
University of Pennsylvania, Philadelphia, PA, USA
e-mail: rsjeong@sas.upenn.edu

A. Keisling
University of Michigan, Ann Arbor, MI, USA
e-mail: keislina@umich.edu

A. Lilly
Reed College, Portland, OR, USA
e-mail: astridlilly@reed.edu

P. Ngamlamai
Harvey Mudd College, Claremont, CA, USA

A. Bahmanian (ed.), *Combinatorics and Combinatorial Computing*, Springer Proceedings in Mathematics & Statistics 516,
https://doi.org/10.1007/978-3-032-08549-8_1

to generalized dihedral groups $D = \mathbb{Z}_2 \ltimes G$, where G is an abelian group of size n and the nonidentity element of $\mathbb{Z}_2$ acts by inversion. We make further progress on the conjecture by considering subsets with a fixed number of rotations and reflections. By bounding the expected number of overlapping sums, we show that the collection $\mathcal{S}_{D,m}$ of subsets of the generalized dihedral group D of size m has more MSTD sets than MDTS sets when $6 \leq m \leq c_j\sqrt{n}$ for $c_j = 1.3229/\sqrt{111+5j}$, where j is the number of elements in G with order at most 2. We also analyze the expectation for $|A+A|$ and $|A-A|$ for $A \subseteq D_{2n}$, proving an explicit formula for $|A-A|$ when n is prime.

Keywords Sumsets · Difference sets · Dihedral groups

MSC Codes: 11P99, 05B10

1 Introduction and Main Results

Given a set of A integers, the sumset and difference set of A are defined as

$$A+A \;=\; \{a_1+a_2 : a_1, a_2 \in A\} \quad \text{and} \quad A-A \;=\; \{a_1-a_2 : a_1, a_2 \in A\}. \tag{1}$$

These elementary operations are fundamental in additive number theory. A natural problem of recent interest has been to understand the relative sizes of the sum and difference sets of sets A.

Definition 1 We say that a set A has **more sums than differences (MSTD)** if $|A+A| > |A-A|$; has **more differences than sums (MDTS)** if $|A+A| < |A-A|$; or is **sum-difference balanced** if $|A+A| = |A-A|$.

We intuitively expect most sets to be MDTS since addition is commutative and subtraction is not. Nevertheless, MSTD subsets of integers exist. Nathanson detailed in [9] the history of the problem, and attributed to John Conway the first recorded example of an MSTD subset of integers, $\{0, 2, 3, 4, 7, 11, 12, 14\}$. Martin and O'Bryant proved in [6] that the proportion of the 2^n subsets A of $\{0, 1, \ldots, n-1\}$ which are MSTD is bounded below by a positive value for all $n \geq 15$. They proved this by controlling the "fringe" elements of A, those close to 0 and $n-1$, which have the most influence over whether elements are missing from the sum and difference sets. In [13], Zhao gave a deterministic algorithm to compute the limit of the ratio of MSTD subsets of $\{0, 1, \ldots, n-1\}$ as n goes to infinity and found that this ratio is at least 4.28×10^{-4}. For more on the problem of MSTD sets in the integers, see also [3] and [10] for constructive examples of infinite families of MSTD sets, [8] and [11] for non-constructive proofs of existence of infinite families of MSTD sets, and [4] and [5] for an analysis of sets with each integer from 0 to $n-1$ included with probability $cn^{-\delta}$.

More recently, several authors have examined analogous problems for groups G other than the integers. For example, Do, Kulkarni, Moon, Wellens, Wilcox, and the seventh author studied in [1] the analogous problem for higher-dimensional integer lattices.

For finite groups, although the usual notation for the operation of the group is multiplication, we match the notation from previous work and define, for a subset $A \subseteq G$, its sumset and difference set as

$$A + A = \{a_1 a_2 : a_1, a_2 \in A\} \quad \text{and} \quad A - A = \{a_1 a_2^{-1} : a_1, a_2 \in A\}. \tag{2}$$

Definition 1 of MSTD, MDTS, and sum-difference balanced sets apply in this context.

The approaches used to study MSTD subsets of integers do not generalize for MSTD subsets of finite groups due to the lack of fringes. Zhao proved asymptotics for numbers of MSTD subsets of finite abelian groups as the size of the group goes to infinity in [12]. The seventh author and Vissuet examined the problem for arbitrary finite groups G, also with the size of the group going to infinity, and proved Theorem 1.

Theorem 1 ([7]) *Let $\{G_n\}$ be a sequence of finite groups, not necessarily abelian, with $|G_n| \to \infty$. Let A_n be a uniformly chosen random subset of G_n. Then $\mathbb{P}[A_n + A_n = A_n - A_n = G_n] \to 1$ as $n \to \infty$. In other words, as the size of the finite groups increases without bound, almost all subsets are balanced (with sumset and difference set equalling the entire group).*

Furthermore, for the case of dihedral groups D_{2n}, they proposed Conjecture 1.

Conjecture 1 ([7]) Let $n \geq 3$ be an integer. There are more MSTD subsets of D_{2n} than MDTS subsets of D_{2n}.

Given a set $A \subseteq D_{2n} = \langle r, s \mid r^n, s^2, rsrs \rangle$, define R (resp. F) as the set of elements of A of the form r^i (resp. $r^i s$), called rotation elements (resp. flip elements). Hence, $A = R \cup F$. Then, we can write

$$\begin{aligned} A + A &= (R + R) \cup (F + F) \cup (R + F) \cup (-R + F), \\ A - A &= (R - R) \cup (F + F) \cup (R + F). \end{aligned} \tag{3}$$

Intuition for Conjecture 1 comes from noting that $F + F$ and $R + F$ contribute to both $A + A$ and $A - A$; $R + R$ and $-R + F$ contribute only to $A + A$; and $R - R$ contributes only to $A - A$.

In 2020, Haviland, Kim, Lâm, Lentfer, Trejos Suáres, and the seventh author made progress towards Conjecture 1 by partitioning subsets of D_{2n} by size. They proposed Conjecture 2 as a means of proving Conjecture 1.

Conjecture 2 ([2]) Let $n \geq 3$ be an integer, and let $\mathcal{S}_{2n,m}$ denote the collection of subsets of D_{2n} of size m. For any $m \leq 2n$, $\mathcal{S}_{2n,m}$ has at least as many MSTD sets as MDTS sets.

Remark 1 Note Lemma 4 implies if $\max(|R|, |F|) > n/2$, then A is not MDTS. This follows from the discussion after the statement of Conjecture 1 (which we explicitly extend to the general dihedral group case in Sect. 3). For a set to be MDTS, $R - R$ must contribute rotations to $A - A$ that the set $A + A$ does not have. But here we have shown that if $\max(|R|, |F|) > n/2$, then $A + A$ has all the rotations.

Lemma 5 *Let D be a generalized dihedral group of size $2n$, and let $A \subseteq D$ with $|A| = m$. If $m > n$, then $A + A = A - A = D$.*

Proof Let R (resp. F) be the subset of rotations (resp. flips) in A; hence $A = R \cup F$. Let L, S be the larger and smaller of R and F, respectively. Define $|L| = n_1, |S| = n_2$ for $n_1 + n_2 = m > n$. Thus, $n_1 > n/2$.

By Lemma 4, we have that $R_D \subseteq A + A$ and $R_D \subseteq A - A$.

For each flip $f \in F_D$, define $fL^{-1} = \{f\ell^{-1} \mid \ell \in L\}$ and $fL = \{f\ell \mid \ell \in L\}$.

Note that $|fL^{-1}| = |fL| = |L| = n_1$, $|S| = n_2$, and fL^{-1}, fL, S are subsets of S_D, which has size $n < n_1 + n_2$. Hence, by the inclusion–exclusion principle, $fL^{-1} \cap S$ and $fL \cap S$ are nonempty. Thus, $f \in S + L \subseteq A + A$ and $f \in S - L \subseteq A - A$. Therefore, $F_D \subseteq A + A$ and $F_D \subseteq A - A$.

Thus, we have $R_D, F_D \subseteq A + A$ and $R_D, F_D \subseteq A - A$, which imply $A + A = A - A = D$. □

3 Collision Analysis

Let G be a finite abelian group of size n, written multiplicatively. Recall that the generalized dihedral group is given by $D = \mathbb{Z}_2 \ltimes G$, where the nonidentity element of $\mathbb{Z}_2$ acts on G by inversion.

This section is dedicated to proving the following.

Theorem 2 Let $D = \mathbb{Z}_2 \ltimes G$ be a generalized dihedral group of size $2n$. Let $\mathcal{S}_{D,m}$ denote the collection of subsets of D of size m, and let j denote the number of elements in G with order at most 2. If $6 \leq m \leq c_j\sqrt{n}$, where $c_j = 1.3229/\sqrt{111 + 5j}$, then there are more MSTD sets than MDTS sets in $\mathcal{S}_{D,m}$.

Note that the theorem is only useful when $c_j\sqrt{n} \geq 6$, or $n \geq (6/c_j)^2$.

If n is arbitrarily large and j is a constant compared to n, we can make a stronger statement: we can replace c_j in the above theorem with a constant arbitrarily close to $\sqrt{2/7} \approx 0.5345$. Specifically, we have the following.

Theorem 4 *For fixed j and $\epsilon > 0$, there exists $n_{j,\epsilon}$ with the following property. Let G be an abelian group of size $n \geq n_{j,\epsilon}$ with at most j elements of order 2 or 1. Then with D and $\mathcal{S}_{D,m}$ defined as in Theorem 2, we have that if $6 \leq m \leq \left(\sqrt{2/7} - \epsilon\right)\sqrt{n}$, then there are more MSTD sets than MDTS sets in $\mathcal{S}_{D,m}$.*

We can give a stronger, more general statement on the proportion of MSTD sets in $\mathcal{S}_{D,m}$ if m is large and also bounded above by a (smaller) constant times $\sqrt{n}$.

Theorem 5 *Let D, j, and $\mathcal{S}_{D,m}$ be defined as in Theorem 2. For any $\epsilon > 0$, there exist m_ϵ and $c_{\epsilon,j}$ such that if $m_\epsilon \leq m \leq c_{\epsilon,j}\sqrt{n}$, the proportion of MSTD sets in $\mathcal{S}_{D,m}$ is at least $1 - \epsilon$.*

Here m_ϵ and $c_{\epsilon,j}$ are independent of n, but similarly to before, this theorem is only useful when $n \geq (m_\epsilon / c_{\epsilon,j})^2$.

Remark 2 In practice, Theorems 2 and 5 are most useful if j is essentially a constant compared to n. This is indeed the case for the original dihedral group $D_{2n} = \mathbb{Z}_2 \ltimes \mathbb{Z}_n$, where $j = 1$ when n is odd and $j = 2$ when n is even, yielding $c_j \geq 0.12$. The family of original dihedral groups is also a good example of how to use Theorem 4: we can apply that theorem with $j = 2$ and ϵ arbitrarily small to get that for large enough dihedral groups, we can get the coefficient of the $\sqrt{n}$ in the theorem to be very close to $\sqrt{2/7}$, which is a significant improvement over 0.12.

However, these theorems are not useful when, for example, $G = \mathbb{Z}_2 \times \ldots \times \mathbb{Z}_2 \times \mathbb{Z}_3$. Here G does have an element of order at least 3, so Conjecture 3 applies, but we have $j = n/3$, which is too large for Theorems 2 and 5 to apply to any values of m. In fact, if $j \geq n/100$, then $n \leq (6/c_j)^2$, and Theorem 2 does not apply to any values of m.

We first prove Theorem 2 here. Then, in Sect. 3.1 we demonstrate Theorems 4 and 5.

Recall the definitions of R_D and F_D from Sect. 2.2. Note that any element in F_D has order 2 in D. Furthermore, any element in R_D has order at most 2 in D if and only if it has order at most 2 in G.

We begin with a set A with size m and count the number of elements in $A + A$ and $A - A$. In this count, we will make a naive assumption: there are no overlaps between sums and differences that we do not expect to overlap. Decompose A into the union of the set of rotations $R = A \cap R_D$ of and the set of flips $F = A \cap F_D$, and define $k = |F|$. We have:

$$\begin{aligned} A + A &= (R + F) \cup (F + R) \cup (R + R) \cup (F + F); \\ A - A &= (R - F) \cup (F - R) \cup (R - R) \cup (F - F) \end{aligned} \tag{5}$$

Consider first the flips in $A + A$ and $A - A$. In $A + A$, these are in $R + F$ and $F + R$. Note that we do not expect a lot of overlap in general; for a rotation $(0, g_1) \in R$ and a flip $(1, g_2) \in F$, $(0, g_1) \cdot (1, g_2) = (1, g_1 g_2)$ does not equal $(1, g_2) \cdot (0, g_1) = (1, g_2 g_1^{-1})$ unless g_1 has order 1 or 2 in G. On the other hand, for the flips in $A - A$, we have $R - F = F - R$. This is because $(0, g_1) \cdot (1, g_2)^{-1} = (0, g_1) \cdot (1, g_2) = (1, g_1 g_2)$, and $(1, g_2) \cdot (0, g_1)^{-1} = (1, g_2) \cdot (0, g_1^{-1}) = (1, g_1 g_2)$, so these two are the same. There are $m - k$ rotations and k flips in A, so we thus expect the flips to contribute $2(m - k)k$ to $A + A$ but only $(m - k)k$ to $A - A$.

Next, consider the rotations in $A + A$ and $A - A$. Begin with $F + F$ and $F - F$. Since all flips have order 2, these are in fact the same set and thus always contribute equally to $A + A$ and to $A - A$. Also note that if $k \neq 0$ (we will treat the $k = 0$ case later), the identity 1 is contained in $F + F$ and $F - F$.

Next consider $R + R$ and $R - R$. Adding rotations is commutative, so we expect $R + R$ to contribute $\binom{m-k}{2} + (m - k)$ to the size of $A + A$, where the $m - k$ term comes from the sum of each rotation in R with itself. On the other hand, in general $g_1 g_2^{-1} \neq g_2 g_1^{-1}$, so $R - R$ is expected to contribute $2\binom{m-k}{2}$. Here there is no additional $m - k$ term since when $g_1 = g_2$, we have $g_1 g_2^{-1} = 1$, and 1 was already counted in $A - A$ from $F - F$.

We now put this all together. For $|A + A| > |A - A|$, we need

$$\begin{aligned} & 2(m-k)k + \binom{m-k}{2} + (m-k) > (m-k)k + 2\binom{m-k}{2}, \\ \Longleftrightarrow\; & k > m/3 - 1 \text{ and } m \neq k, \\ \Longleftrightarrow\; & m/3 \leq k < m. \end{aligned} \tag{6}$$

Note that when $k = m$ the set is necessarily balanced as $F + F = F - F$. Further, one can now see why we may assume $k \neq 0$: when k is smaller than $m/3$, we expect the set to be MDTS, and indeed we will assume this is the case.

To use this naive estimate to prove our theorem, we first formalize our assumption that we have minimal overlaps within the sumset.

Definition 2 Let $A \subseteq D$ such that $|A| = m$, and let $q = (a, b, c, d) \in A^4$. We say that q represents a **collision** if $ab = cd$.

Every collision that occurs in A has the potential to make $|A+A|$ smaller relative to our naive estimate, unless the quadruple q is of a form that we already took into account. For example, if $q = (a, b, a, b)$, then this is not a collision we need to count as $ab = ab$ trivially. Similarly, collisions of the form $q = (a, b, b, a)$ with a and b both rotations do not subtract from $|A + A|$ in Eq. (6) as we already accounted for commutativity of addition for rotations. And finally, if $q = (a, b, c, d)$ is a collision where a, b, c, and d are all flips, then q does not impact Eq. (6) as $F + F = F - F$ do not affect the relative sizes of the sum and difference sets. We refer to these three kinds of quadruples as **redundant**.

Every non-redundant collision (a, b, c, d) of A, together with (c, d, a, b), decreases the size of $A + A$ by at most 1 from our naive estimate. Let X_A be half the total number of non-redundant collisions of A. Then combining the above analysis with Eq. (6), we are guaranteed to have that A is MSTD when

$$\begin{aligned} & 2(m-k)k + \binom{m-k}{2} + (m-k) - X_A > (m-k)k + 2\binom{m-k}{2}, \\ \Longleftrightarrow\; & m/3 + \frac{2X_A}{3(m-k)} \leq k < m, \\ \Longleftrightarrow\; & 3k^2 - 4mk + m^2 + 2X_A \leq 0 \text{ (and } k < m). \end{aligned} \tag{7}$$

We use the quadratic equation to solve for when the above quantity equals 0 and obtain $k = (1/6)(4m \pm \sqrt{16m^2 - 12(m^2 + 2X_A)}) = (1/3)(2m \pm \sqrt{m^2 - 6X_A})$. Thus Eq. (7) is satisfied when:

$$\frac{2m - \sqrt{m^2 - 6X_A}}{3} \leq k \leq \frac{2m + \sqrt{m^2 - 6X_A}}{3} \quad \text{(and } k < m\text{)}. \tag{8}$$

Remark 3 We are assuming that no "collisions" of the form $ab^{-1} = cd^{-1}$ happen to lower the size of $A - A$. Because our objective is to guarantee $|A + A| > |A - A|$ for a large proportion of A, this assumption still gives a sufficient condition on X_A and k.

One therefore sees that for most values of k, when the number of collisions is not too large, A is MSTD. More formally, suppose that A is chosen randomly out of the subsets of D with size m. Suppose that the expectation value of X_A is bounded above by $c_1 m^2$, where $c_1 = 7/1152 \approx 0.006076$. Then the actual value of X_A exceeds $12c_1 m^2$ at most $1/12$ of the time by Markov's inequality. Of sets A with $5m/12 \leq k \leq 11m/12$, the actual value of X_A exceeds $12c_1 m^2$ at most $2/12 = 1/6$ of the time. Thus when $5m/12 \leq k \leq 11m/12$, Eq. (8) is true at least $5/6$ of the time since when $X_A = 12c_1 m^2$, the equation reads

$$\begin{aligned} &\frac{2m - \sqrt{m^2 - 72c_1 m^2}}{3} \leq k \leq \frac{2m + \sqrt{m^2 - 72c_1 m^2}}{3}, \\ \Longleftrightarrow\ &\frac{2m - \sqrt{9/16 \cdot m^2}}{3} \leq k \leq \frac{2m + \sqrt{9/16 \cdot m^2}}{3}, \\ \Longleftrightarrow\ &\frac{5m}{12} \leq k \leq \frac{11m}{12}. \end{aligned} \tag{9}$$

Now, we need to make sure that for our values of m, more than $(1/2)/(5/6) = 3/5$ proportion of sets in $\mathcal{S}_{D,m}$ have $5m/12 \leq k \leq 11m/12$. This will ensure that a proportion greater than $1/2$ of sets in $\mathcal{S}_{D,m}$ satisfy Eq. (8) and are therefore MSTD.

More formally speaking, we require

$$\lim_{n\to\infty} \frac{\sum_{k=\lceil 5m/12 \rceil}^{\lfloor 11m/12 \rfloor} \binom{n}{k}\binom{n}{m-k}}{\binom{2n}{m}} > 3/5 \tag{10}$$

$$\Longleftrightarrow \frac{m!}{2^m} \sum_{k=\lceil 5m/12 \rceil}^{\lfloor 11m/12 \rfloor} \frac{1}{k!(m-k)!} > 3/5, \tag{11}$$

$$\mathbb{P}[C \mid B] \geq 1 - \frac{1 - ((1/2)/(1-\epsilon_1) + \epsilon_1)}{1-\epsilon_1} \tag{18}$$

$$= 1 - \frac{(1-\epsilon_1) - (1/2)/(1-\epsilon_1)}{1-\epsilon_1} \tag{19}$$

$$= \frac{1/2}{(1-\epsilon_1)^2}, \tag{20}$$

and the claim is shown.

To ensure that Eq. (16) is true, we require

$$\mathbb{E}[X_A] \leq \left(1 - \epsilon_1 - \frac{1/2}{1-\epsilon_1}\right)\left(\frac{m^2}{8} - \frac{m\epsilon_1}{6} - \frac{\epsilon_1^2}{6}\right). \tag{21}$$

Then by Markov's inequality, the probability that X_A exceeds $\frac{m^2}{8} - \frac{m\epsilon_1}{6} - \frac{\epsilon_1^2}{6}$ is at most $1 - \epsilon_1 - \frac{1/2}{1-\epsilon_1}$, which is equivalent to Eq. (16).

We may now choose a small value ϵ_2 such that Eq. (21) is true if

$$\mathbb{E}[X_A] \leq \left(\frac{1}{16} - \epsilon_2\right) m^2. \tag{22}$$

Notice that in the limit $\epsilon_1 \to 0$, Eq. (21) boils down to the statement that $\mathbb{E}[X_A] \leq m^2/16$, so we can make ϵ_2 be as small as desired by making ϵ_1 be small.

We now revisit Lemma 6. We are now assuming that m is large and j is a constant, so only the first term dominates:

$$\mathbb{E}[X_A] \leq \left(\frac{7}{32} + \epsilon_3\right) \frac{m^4}{n}. \tag{23}$$

We wanted to have $\mathbb{E}[X_A] \leq (1/16 - \epsilon_2)m^2$. Thus the upper bound on m is now determined as

$$\left(\frac{7}{32} + \epsilon_3\right) \frac{m^4}{n} \leq \left(\frac{1}{16} - \epsilon_2\right) m^2$$

$$\iff m \leq c\sqrt{n}, \tag{24}$$

where

$$c = \sqrt{\frac{1/16 - \epsilon_2}{7/32 + \epsilon_3}}. \tag{25}$$

If n is arbitrarily large and j is constant compared to n, we can choose ϵ_2 and ϵ_3 to be very small, so that c is arbitrarily close to $\sqrt{(1/16)/(7/32)} \approx 0.5345$.

This completes the proof of Theorem 4. To prove Theorem 5, we follow a very similar method. We choose m_ϵ large enough that almost all of the sets in $\mathcal{S}_{D,m}$ have $(0.5 - \epsilon_1)m \leq k \leq (0.5 + \epsilon_1)m$. The only difference is that now, in Eq. (15) we replace the right-hand side with $(1 - \epsilon_1)$, so that the proportion of MSTD sets is at least $(1 - \epsilon_1)^2$. (We may choose ϵ_1 so that $(1 - \epsilon_1)^2$ equals the desired $(1 - \epsilon)$). Then in Eq. (16) we replace the right-hand side with $1 - (1 - \epsilon_1)\epsilon_1$, leading to the analog of Eq. (22) being that $\mathbb{E}[X_A] \leq \epsilon_2 m^2$ for some small ϵ_2 depending on ϵ_1. The rest of the proof continues as before, leading to a coefficient $c_{\epsilon,j}$ proportional to $\sqrt{\epsilon_2}$. This completes the proof.

3.2 Proof of Lemma 6

We now prove Lemma 6. Recall that X_A is defined to be half the number of non-redundant collisions in the set A, and we are interested in bounding above the expectation value of X_A when A is chosen uniformly at random from $\mathcal{S}_{D,m}$.

To more easily count the collisions in A, we make the following definition.

Definition 3 A **redundant triple** is a triple $(a, b, c) \in D^3$ such that the quadruple $(a, b, c, c^{-1}ab)$ is redundant. That is, a triple (a, b, c) is redundant if $a = c$, or if a, b, and c are all flips, or if $b = c$ and a and b are both rotations. Denote $T \subseteq D^3$ to be the set of non-redundant triples.

Define the function $\chi : \mathcal{S}_{D,m} \times T \to \{0, 1\}$ by $\chi(A, t) = 1$ if for the non-redundant triple $t = (a, b, c)$, the element $c^{-1}ab$ is in A, and $\chi(A, t) = 0$ otherwise.

For a fixed set A, the set $A^3 \cap T$ is the set of non-redundant triples with all three elements contained in A. Notice that we have

$$X_A = \frac{1}{2} \sum_{t \in A^3 \cap T} \chi(A, t). \tag{26}$$

That is, the number of non-redundant collisions in A is the same as the number of non-redundant triples $(a, b, c) \in A^3 \cap T$ such that the element $d = c^{-1}ab$ is in A, forming a quadruple (a, b, c, d) representing a collision $ab = cd$.

By definition of expectation value, we write

$$\mathbb{E}[X_A] = \sum_{A \in \mathcal{S}_{D,m}} \mathbb{P}[A] X_A. \tag{27}$$

Since A is chosen uniformly at random from the $\binom{2n}{m}$ sets in $\mathcal{S}_{D,m}$, we have $\mathbb{P}[A] = 1/\binom{2n}{m}$. Thus,

$$\mathbb{E}[X_A] = \frac{1}{\binom{2n}{m}} \sum_{A \in \mathcal{S}_{D,m}} \frac{1}{2} \sum_{t \in A^3 \cap T} \chi(A, t). \tag{28}$$

We swap the order of the sums.

$$\mathbb{E}[X_A] = \frac{1}{2}\frac{1}{\binom{2n}{m}} \sum_{t\in T} \sum_{\substack{A\in\mathcal{S}_{D,m}\\ A^3\ni t}} \chi(A,t). \tag{29}$$

To compute the inner sum, we must simply count the number of sets $A \in \mathcal{S}_{D,m}$ with $t = (a, b, c) \in A^3$ such that A contains $c^{-1}ab$. That is,

$$\mathbb{E}[X_A] = \frac{1}{2}\frac{1}{\binom{2n}{m}} \sum_{(a,b,c)\in T} |\{A \in \mathcal{S}_{D,m} \mid a, b, c, c^{-1}ab \in A\}|. \tag{30}$$

We now break this sum into seven pieces for different kinds of triples $(a, b, c) \in T$. These are:

- $T_1 = \{(a, b, c) \in T \mid a, b, c, c^{-1}ab$ are distinct $\}$
- $T_2 = \{(a, b, c) \in T \mid a = b;\ a, c, c^{-1}a^2$ are distinct$\}$
- $T_3 = \{(a, b, c) \in T \mid b = c;\ a, c, c^{-1}ac$ are distinct$\}$
- $T_4 = \{(a, b, c) \in T \mid c^{-1}ab = a;\ a, b, c$ are distinct$\}$
- $T_5 = \{(a, b, c) \in T \mid c^{-1}ab = c;\ a, b, c$ are distinct$\}$
- $T_6 = \{(a, b, c) \in T \mid b = c; a = c^{-1}ac;\ a, c$ are distinct$\}$
- $T_7 = \{(a, b, c) \in T \mid a = b; c = c^{-1}a^2;\ a, c$ are distinct$\}$

We have $T = \bigcup_{i=1}^{7} T_i$, for these seven cases cover all the cases of possible equalities between the four elements except for those where $a = c$, or equivalently, $c^{-1}ab = b$, since those cases are redundant triples. Furthermore, this union is disjoint.

Note that for triples $(a, b, c) \in T_1$, the quantity $|\{A \in \mathcal{S}_{D,m} \mid a, b, c, c^{-1}ab \in A\}|$ is given by $\binom{2n-4}{m-4}$ since we are requiring four distinct elements to be in A, and we have $2n - 4$ choices for the remaining $m - 4$ elements. For triples in T_2, T_3, T_4, and T_5, we are requiring three distinct elements to be in A, so we have $|\{A \in \mathcal{S}_{D,m} \mid a, b, c, c^{-1}ab \in A\}| = \binom{2n-3}{m-3}$. Finally, for triples in T_6 and T_7 we have $|\{A \in \mathcal{S}_{D,m} \mid a, b, c, c^{-1}ab \in A\}| = \binom{2n-2}{m-2}$.

Therefore, from Eq. (30), letting $t_{2\text{—}5} = |T_2| + |T_3| + |T_4| + |T_5|$, we may write:

$$\begin{aligned}\mathbb{E}[X_A] &= \frac{1}{2}\frac{1}{\binom{2n}{m}}\left[\binom{2n-4}{m-4}|T_1| + \binom{2n-3}{m-3}(t_{2\text{—}5}) + \binom{2n-2}{m-2}(|T_6| + |T_7|)\right]\\ &\le \frac{1}{2}\left[\left(\frac{m}{2n}\right)^4|T_1| + \left(\frac{m}{2n}\right)^3(t_{2\text{—}5}) + \left(\frac{m}{2n}\right)^2(|T_6| + |T_7|)\right],\end{aligned} \tag{31}$$

where in the second line we used the fact that

$$\frac{\binom{2n-4}{m-4}}{\binom{2n}{m}} = \frac{m(m-1)(m-2)(m-3)}{2n(2n-1)(2n-2)(2n-3)} \le \left(\frac{m}{2n}\right)^4, \tag{32}$$

and similarly for the other two terms.

Now, to use Eq. (31) to find an upper bound on $\mathbb{E}[X_A]$, all that remains to be done is find an upper bound on each of the $|T_i|$'s. We do so next.

We bound $|T_1|$ using the trivial inequality $|T_1| \leq |T|$. There are $(2n)^3$ total triples in D^3, but we may subtract the redundant triples, including the n^3 triples consisting of three flips. Thus we obtain

$$|T_1| \leq 7n^3. \tag{33}$$

We bound $|T_2|$ and $|T_3|$ next. We have $2n$ choices for b. In T_2, a must equal b, and in T_3, c must equal b, and in both, we have at most $2n$ choices for the remaining value of a or c. The extra condition that $c^{-1}ab$ is distinct from the others only lowers $|T_2|$ and $|T_3|$, so we do not have to take it into account to obtain an upper bound. Thus,

$$|T_2|,\ |T_3| \leq 4n^2. \tag{34}$$

For $|T_4|$ and $|T_5|$, we have $2n$ choices for a and $2n$ choices for c, but b must be the element $a^{-1}ca$ for T_4 or $a^{-1}c^2$ for T_5. Thus we have

$$|T_4|,\ |T_5| \leq 4n^2. \tag{35}$$

When considering T_6, we note that since $b = c$, the triple (a, b, c) is redundant if a and c are both rotations, or if both are flips. So, one must be a rotation and the other must be a flip, and we require that $a = c^{-1}ac$, or $ca = ac$. Thus to bound $|T_6|$ we must count the number of pairs of elements with $(0, g_1) \cdot (1, g_2) = (1, g_2) \cdot (0, g_1)$, that is, $(1, g_1g_2) = (1, g_2g_1^{-1})$. This happens if and only if $g_1^{-1} = g_1$, or $g_1^2 = 1$. Recalling that there are j elements in G with order 2 or less, there are therefore j choices for the rotation element, and n choices for the flip element. We multiply by 2 since a can be either the flip or the rotation, and c is the other of the two. Thus,

$$|T_6| \leq 2nj. \tag{36}$$

Finally, for T_7, we again split into two cases: firstly where a is a rotation, and secondly where a is a flip, so c is a rotation to avoid redundancy. Here we require $c = c^{-1}a^2$, or $c^2 = a^2$. For the first case, we first consider the number of pairs with $a^2 = c^2 = 1$. Since a must be a rotation, there are only j choices for $a^2 = 1$; c can be a flip or a rotation, so there are $n + j$ ways to have $c^2 = 1$. So, there are at most $(n + j)j$ pairs of this kind. If a is a rotation and $a^2 \neq 1$, then c must also be a rotation since otherwise $c^2 = 1 \neq a^2$. Thus a and c commute, so $a^2 = c^2 \iff (ac^{-1})^2 = 1$. There are $n - j$ choices of a with $a^2 \neq 1$, and for each, there are j choices of c which have $(ac^{-1})^2 = 1$. Thus there are at most $(n - j)j$ pairs of this kind.

Next we consider the case where a is a flip and c is a rotation. Here $a^2 = 1$, so there are only j choices for c that have $c^2 = a^2$. Thus there are nj pairs of this kind. In total,

$$|T_7| \leq (n + j)j + (n - j)j + nj = 3nj. \tag{37}$$

We present an example of the number of collisions in another dihedral group $\mathbb{Z}_2 \ltimes \mathbb{Z}_n^2$. We show that the dihedral group $\mathbb{Z}_2 \ltimes \mathbb{Z}_n^2$ possesses exactly the same number of possible collisions as the dihedral group $\mathbb{Z}_2 \ltimes \mathbb{Z}_{n^2}$ if and only if n is odd, and there are more collisions in $\mathbb{Z}_2 \ltimes \mathbb{Z}_n^2$ otherwise. This result only provides some intuition on the expected number of collisions depending on j, the number of elements of order 2 within the particular abelian group.

Lemma 7 *The number of possible collisions within the two groups are equal if and only if n is odd, and there are more collisions in $\mathbb{Z}_2 \ltimes \mathbb{Z}_n^2$ otherwise.*

To do so, we make use of the following useful results:

Lemma 8 *There are n^2 pairs $(a_1, a_2), (b_1, b_2) \in \mathbb{Z}_n^2$ where $a_1 - b_1 \equiv x \mod n$ and $a_2 - b_2 \equiv y \mod n$. There are also n^2 pairs $a_1 n + a_2, b_1 n + b_2 \in \mathbb{Z}_{n^2}$ where $(a_1 n + a_2) - (b_1 n + b_2) \equiv xn + y \mod n^2$.*

Proof For any given (b_1, b_2) and $b_1 n + b_2$, there exist only a single (a_1, a_2) and $a_1 n + a_2$ which satisfies the conditions respectively. Thus, there are n^2 such pairs for both cases. □

Lemma 9 *The number of pairs $(a_1, a_2), (b_1, b_2) \in \mathbb{Z}_n^2$ where $a_1 + b_1 \equiv x \mod n$ and $a_2 + b_2 \equiv y \mod n$ is as follows:*

- *when n is odd, there are $\frac{n^2+1}{2}$ such pairs*
- *when n is even and y is odd or x is odd, there are $\frac{n^2}{2}$ such pairs*
- *when n, y, and x are all even, there are $\frac{n^2+4}{2}$ such pairs.*

Proof When n is odd, each choice of a_1 is paired with a single possible b_1, with one such pair being $a_1 = b_1$. Similarly, the same holds for a_2 and b_2. This gives n^2 as the number of pairs. However, we over-counted since swapping a_1 with b_1 and a_2 with b_2 does not yield a distinct pair. Thus, we divide by 2 except for the single pair where $a_1 = b_1$ and $a_2 = b_2$ which we did not over-count to get $\frac{n^2-1}{2} + 1 = \frac{n^2+1}{2}$ pairs of $(a_1, a_2), (b_1, b_2)$.

When n is even but y is odd, we choose pairs (a_2, b_2) first which gives $n/2$ distinct pairs where a_2 is never equal to b_2. For each of these pairs, we can choose any choice of a_1 which forces b_1 without over-counting. Thus, we get $\frac{n^2}{2}$ pairs. Similar arguments hold for when x is odd.

When all of n, y, and x are even, we get that there are two pairs of identical number and $\frac{n-2}{2}$ pairs of different numbers which sum to x and similar for y in $\mathbb{Z}_n$. We choose a_2, b_2 first. If $a_2 = b_2$, we have $\frac{n-2}{2} + 2 = \frac{n+2}{2}$ choices for choosing a_1, b_1. If $a_2 \neq b_2$, we have that for any a_1, we uniquely determines b_1 without over-counting and thus there are n choices. In total, we have $2 * \frac{n+2}{2} + n * \frac{n-2}{2} = \frac{n^2+4}{2}$ pairs of $(a_1, a_2), (b_1, b_2)$. □

Lemma 10 *The number of pairs $a_1 n + a_2, b_1 n + b_2 \in \mathbb{Z}_{n^2}$ where $(a_1 n + a_2) + (b_1 n + b_2) \equiv xn + y \mod n^2$ is as follows:*

- *when n is odd, there are $\frac{n^2+1}{2}$ such pairs*

- *when n is even but y is odd, there are* $\frac{n^2}{2}$ *such pairs*
- *when both n and y are even, there are* $\frac{n^2+2}{2}$ *such pairs*

Proof When n is odd, we have that each of a_1n+a_2 is paired with another b_1n+b_2, with exactly one being paired with itself. Thus, there are $\frac{n^2-1}{2}+1=\frac{n^2+1}{2}$ pairs of a_1n+a_2 and b_1n+b_2.

When n is even but y is odd, we have that each of a_1n+a_2 is paired with another b_1n+b_2, necessarily distinct. Thus, there are $\frac{n^2}{2}$ pairs of a_1n+a_2 and b_1n+b_2.

When n is even and y is even, we have that each of a_1n+a_2 is paired with another b_1n+b_2, with exactly two being paired with themselves. Thus, there are $\frac{n^2-2}{2}+2=\frac{n^2+2}{2}$ pairs of a_1n+a_2 and b_1n+b_2. □

Combining these results over the casework where elements of our pairs may be a rotation or a flip yield the desired result, with the details of the proof in Appendix 2.

4 Expected Size of Sum and Difference Sets

In this section, we only consider the classical dihedral groups

$$D_{2n}=\mathbb{Z}_2\ltimes\mathbb{Z}_n=\langle r,s\mid r^n,s^2,rsrs\rangle \tag{44}$$

Throughout this section, we use $\mathcal{S}_{2n,m}$ to denote the set of subsets of size m in D_{2n}.

The method of collision analysis will likely not be sufficient to prove that $\mathcal{S}_{2n,m}$ has more MSTD sets than MDTS sets for values of m greater in order of magnitude than $\sqrt{n}$. The intuition for this comes from the fact that the sum and difference sets for $A\subseteq D_{2n}$ should very roughly have size on the same order of magnitude as A^2. Hence, one would expect to usually have $A+A=A-A=D_{2n}$ when m is much greater than $\sqrt{n}$. The analysis for relative numbers of MSTD and MDTS sets in $\mathcal{S}_{2n,m}$ for these larger values of m should therefore be based on counting the number of *missed* sums and differences in D_{2n}, in direct analogy with the case of slow decay for the integers in [4].

We take the first steps toward such an analysis by proving the following special case.

Theorem 3 If n is prime, and A is chosen uniformly at random from $\mathcal{S}_{2n,m}$, then $\mathbb{E}[|A-A|]$ is given by

$$2n-\frac{nm2^m\binom{n}{m}+2n(n-1)\binom{n-m-1}{m-1}}{m\binom{2n}{m}}-\frac{n^2(n-1)}{\binom{2n}{m}}\sum_{k=1}^{m-1}\frac{\binom{n+k-m-1}{m-k-1}\binom{n-k-1}{k-1}}{k(m-k)}. \tag{45}$$

The behavior of this expectation depends drastically on the size of m compared to n; see Fig. 1 on page 23. Theorem 3 follows from the following straightforward yet useful lemma, which reduces the problem of computing the probability of missing a sum or difference to an analogous problem in $\mathbb{Z}_n$.

Lemma 11 *Let A be a subset in $\mathcal{S}_{2n,m}$ chosen uniformly at random. For $0 \leq k \leq m$, define the following probabilities:*

$$\begin{aligned} p_1(k) &= \mathbb{P}[A \text{ has } k \text{ flips}], \\ p_2(k) &= \mathbb{P}[i \notin S + S | S \subseteq \mathbb{Z}_n, |S| = k], \\ p_3(k) &= \mathbb{P}[i \notin S - S | S \subseteq \mathbb{Z}_n, |S| = k], \\ p_4(k) &= \mathbb{P}[i \notin S_1 + S_2 \wedge i \notin S_2 - S_1 | S_1, S_2 \subseteq \mathbb{Z}_n, |S_1| = m-k, |S_2| = k], \text{ and} \\ p_5(k) &= \mathbb{P}[i \notin S_1 + S_2 | S_1, S_2 \subseteq \mathbb{Z}_n, |S_1| = m-k, |S_2| = k]. \end{aligned} \tag{46}$$

Then if r^i is a rotation in D_{2n}, we have

$$\mathbb{P}[r^i \notin A + A] = \sum_{k=0}^{m} p_1(k) \cdot p_2(m-k) \cdot p_3(k), \text{ and} \tag{47}$$

$$\mathbb{P}[r^i \notin A - A] = \sum_{k=0}^{m} p_1(k) \cdot p_3(m-k) \cdot p_3(k). \tag{48}$$

If $r^i s$ is a flip in D_{2n}, we have

$$\mathbb{P}[r^i s \notin A + A] = \sum_{k=0}^{m} p_1(k) \cdot p_4(k), \text{ and} \tag{49}$$

$$\mathbb{P}[r^i s \notin A - A] = \sum_{k=0}^{m} p_1(k) \cdot p_5(k). \tag{50}$$

Proof Partition A into its set of rotations R and flips F, and suppose that $|F| = k$ and $|R| = m - k$. The rotation element r^i can appear in $A + A$ precisely as $r^j r^\ell = r^{j+\ell}$ for $r^j, r^\ell \in R$ or as $(r^j s)(r^\ell s) = r^{j-\ell}$ for $r^j s, r^\ell s \in F$. Taking the probability of the negations of these events respectively give second and third probabilities appearing in the sum of Eq. (47). Equation (48) follows similarly.

The flip element $r^i s$ can appear in $A + A$ precisely as $r^j (r^\ell s) = r^{j+\ell} s$ or as $(r^\ell s) r^j = r^{\ell - j} s$ for $r^j \in R$, $r^\ell s \in F$, but unlike in the previous cases, these events are no longer independent so Eq. (49) cannot be broken up into a product of simpler probabilities. Finally, $r^i s$ can appear in $A - A$ precisely $r^j (r^\ell s)^{-1} = r^{j+\ell} s = (r^\ell s)(r^j)^{-1}$, from which Eq. (50) follows. □

A number of these probabilities can be expressed explicitly in terms of n, m, and k. To prove Theorem 3, we need to compute all probabilities appearing in

Eqs. (48) and (50), but we will also compute the probabilities appearing in Eq. (47) for completeness.

Lemma 12 *Let S be a subset of $\mathbb{Z}_n$ of size $m-k$ chosen uniformly at random, and let i be any element of $\mathbb{Z}_n$. Then*

$$\mathbb{P}[i \notin S+S] = \begin{cases} \frac{2^{m-k}\binom{\frac{n}{2}-1}{m-k}}{\binom{n}{m-k}}, & n \text{ and } i \text{ both even}, \\ \frac{2^{m-k}\binom{\frac{n}{2}}{m-k}}{\binom{n}{m-k}}, & n \text{ even and } i \text{ odd}, \\ \frac{2^{m-k}\binom{\frac{n-1}{2}}{m-k}}{\binom{n}{m-k}}, & n \text{ odd}. \end{cases} \tag{51}$$

Proof If n and i are both even, then there exist exactly 2 elements of $\mathbb{Z}_n$ that give i when added to themselves. The remaining elements of $\mathbb{Z}_n$ partition into pairs of distinct elements adding to i, and any S such that $i \notin S+S$ is obtained by selecting $m-k$ of these $\frac{n}{2}-1$ pairs and one element from each pair, hence the result. The case where n is even and i is odd is identical except that all n of the elements of $\mathbb{Z}_n$ are now partitioned into pairs as there are no elements that give i when added to themselves. Finally, if n is odd, then there is always a unique element that gives i when added to itself, and the remaining elements partition into $\frac{n-1}{2}$ pairs, after which the same analysis can be applied. □

Lemma 13 *Let S be a subset of $\mathbb{Z}_n$ of size k chosen uniformly at random, and let i be any nonzero element of $\mathbb{Z}_n$ of order n/d. Then*

$$\mathbb{P}[i \notin S-S] = \frac{1}{\binom{n}{k}} \sum_{\substack{(k_1,\ldots,k_d)\in\mathbb{Z}_{\geq 0}^d \\ k_1+\cdots+k_d=k}} \prod_{t=1}^{d} g\left(\frac{n}{d}, k_t\right), \tag{52}$$

where

$$g\left(\frac{n}{d}, k_t\right) = \begin{cases} 1, & k_t = 0, \\ \frac{\frac{n}{d}\binom{\frac{n}{d}-1-k_t}{k_t-1}}{k_t}, & k_t > 0. \end{cases} \tag{53}$$

Remark 4 When n is prime, we must have $d = 1$ always and Eq. (52) simplifies significantly to

$$\mathbb{P}[i \notin S-S] = \frac{g(n,k)}{\binom{n}{k}} = \frac{1}{\binom{n}{k}} \cdot \begin{cases} 1, & k = 0, \\ \frac{n\binom{n-1-k}{k-1}}{k}, & k > 0. \end{cases} \tag{54}$$

This is the reason why we restrict to such n in Theorem 3.

Appendix 1: Proof of Lemmas 2 and 3

Lemma 2 **([2])** Let $n \geq 3$, and let $\mathcal{S}_{2n,2}$ denote the collection of subsets of D_{2n} of size 2. Then, $\mathcal{S}_{2n,2}$ has strictly more MSTD sets than MDTS sets.

Proof There are only 3 possible cases to consider for A.

- A contains two flip elements: in this case, adding and subtracting flips is an identical operation as each flip element has order 2. Thus, A will necessarily be sum-difference balanced.
- A contains one flip and one rotation element: let $r^i, r^j s \in A$. Here, $A + A = \{1, r^{2i}, r^{i+j}s, r^{j-i}s\}$. However, $A - A = \{1, r^{i+j}s\}$. When $r^{2i} \neq e$ or $r^{j-i}s \neq r^{i+j}s$, A will be MSTD. In the case where both of these are actually equalities $(A + A = \{e, r^{i+j}s\})$, A will simply be balanced.
- A contains two rotation elements: let $r^a, r^b \in A$. Here, $A + A = \{r^{2i}, r^{i+j}, r^{2j}\}$. However, $A - A = \{1, r^{i-j}, r^{j-i}\}$. Note that $i \neq j$. Both the sum set and the difference set have 3 elements except in one special case. Suppose that $r^{2i} = r^{2j}$. Then, we have that $r^i r^{-j} = r^j r^{-i} \implies r^{i-j} = r^{j-i}$. Thus, when A contains two rotation elements only, A is always balanced.

Therefore, A has strictly more MSTD sets in $\mathcal{S}_{2n,2}$ than MDTS sets. □

Lemma 3 Let $n \geq 3$, and let $\mathcal{S}_{2n,3}$ denote the collection of subsets of D_{2n} of size 3. Then, $\mathcal{S}_{2n,3}$ has strictly more MSTD sets than MDTS sets.

Proof There are 4 possible cases to consider for A.

- A contains three flip elements: just as in the $m = 2$ case, since addition and subtraction of two flips are equivalent, A is balanced.
- A contains two flip elements and one rotation element: let $A = \{r^i, r^j s, r^k s\}$ for $j \neq k$. Then the sumset is

$$A + A = \{r^{2i}, r^{i+j}s, r^{i+k}s, r^{j-k}, r^{k-j}, 1, r^{j-i}s, r^{k-i}s\}, \tag{57}$$

  while the difference set is

$$A - A = \{r^{i+j}s, r^{i+k}s, r^{j-k}, r^{k-j}, 1\}, \tag{58}$$

  so $A - A \subseteq A + A$. Moreover, A is MSTD precisely when at least one of $r^{j-i}s$ and $r^{k-i}s$ is distinct from each of the flips in $A + A$ that lie in $A - A$. We enumerate all the ways that this could fail to occur:

  We have $r^{i+j}s = r^{j-i}s$ and $r^{i+k}s = r^{k-i}s$ if and only if $i = -i \pmod{n}$. This happens if and only if $i = 0$ or if n is even and $i = n/2$, and in each of these cases we have $\binom{n}{2} = \frac{1}{2}(n^2 - n)$ choices for j and k. On the other hand, we have $r^{i+j}s = r^{k-i}s$ and $r^{i+k}s = r^{j-i}s$ if and only if $2i = k - j = j - k \pmod{n}$, which implies that $2j = 2k \pmod{n}$. We also require $2i = 1 \pmod{n}$; because $j \neq k$, these equations can occur if and only if n is divisible by 4, $i = n/4$ or

$3n/4$, and $k - j = n/2 \pmod n$. We therefore have 2 choices for i, and $\frac{1}{2}n$ choices for j and one corresponding choice of k for each i. Finally, note that we cannot have $r^{j-i}s = r^{k-i}s$ because $k \neq i$, so this completes the enumeration of all such sets A that fail to be MSTD.

There are $n\binom{n}{2} = \frac{1}{2}(n^3 - n^2)$ total sets A containing exactly 2 flips. Subtracting the balanced sets that we just enumerated, we obtain the following numbers of MSTD sets:

$$\begin{aligned} \frac{1}{2}(n^3 - n^2) \text{ if } n = 1 \pmod 2, \\ \frac{1}{2}(n^3 - n^2) - n(n-1) \text{ if } n = 2 \pmod 4, \\ \frac{1}{2}(n^3 - n^2) - n(n-1) - n \text{ if } n = 0 \pmod 4. \end{aligned} \tag{59}$$

- A contains one flip element and two rotation element: let $A = \{r^i, r^j, r^k s\}$ for $i \neq j$. Then the sumset is

$$A + A = \{r^{i+j}, r^{2i}, r^{2j}, r^{i+k}s, r^{j+k}s, r^{k-i}s, r^{k-j}s, 1\}, \tag{60}$$

while the difference set is

$$A - A = \{r^{i-j}, r^{j-i}, r^{i+k}s, r^{j+k}s, 1\}, \tag{61}$$

We show that A cannot be MDTS by separately comparing the flips and the rotations in $A + A$ and $A - A$. The flips in $A - A$ are contained in $A + A$, so this comparison is trivial. The rotations in $A - A$ are r^{i-j}, r^{j-i}, and 1; we will show that $A + A$ has at least as many rotations. Consider the rotations in $A + A$: note that r^{i+j} is different from r^{2i} and from r^{2j} since $i \neq j$, so there are at least 2 distinct rotations in $A + A$. If $r^{2i} \neq r^{2j}$, then in fact $A + A$ has at least three distinct rotations, and we have $|A + A| \geq |A - A|$. On the other hand, if $r^{2i} = r^{2j}$, then $r^{i-j} = r^{j-i}$, so there are only at most 2 distinct rotations in $A - A$, so again $|A + A| \geq |A - A|$.
- A contains three rotation elements: for this case, there are $\binom{n}{3}$ such subsets. We compare directly to Eq. (59), the case that yields the least possible number of MSTD sets in $\mathcal{S}_{2n,3}$. Here we have

$$\frac{1}{2}(n^3 - n^2) - n(n-1) - n > \frac{1}{6}n(n-1)(n-2) \tag{62}$$

or equivalently,

$$n^3 - 3n^2 - n > 0, \tag{63}$$

which is true if $n \geq 4$. When $n = 3$, since $n = 1 \mod 2$, we compare $\binom{n}{3}$ to $\frac{1}{2}(n^3 - n^2)$ instead, and the result is verified once again.

Therefore, even if all of the subsets with three rotation elements are MDTS, which is never true to begin with, we still have strictly more MSTD subsets of size 3 than MDTS subsets of size 3. □

Appendix 2: Proof of Lemma 7

For any given $(x, y, F) \in \mathbb{Z}_2 \ltimes \mathbb{Z}_n^2$, we calculate the number of pairs $a, b \in \mathbb{Z}_2 \ltimes \mathbb{Z}_n^2$ where $a - b = (x, y, F)$. Similarly, we also calculate the number of pairs $a, b \in \mathbb{Z}_2 \ltimes \mathbb{Z}_n^2$ where $a + b = (x, y, F)$. Note that these can be counted through counting the number of pairs in $\mathbb{Z}_n^2$ which results in (x, y) as follows:

Case 1, $F = 0$

- Case 1.1, $a - b = (x, y, 0)$
  - Case 1.1.1 a and b are not flips.
    We get that this is equivalent to counting pairs $(a_1, a_2), (b_1, b_2) \in \mathbb{Z}_n^2$ where $a_1 - b_1 \equiv x \mod n$ and $a_2 - b_2 \equiv y \mod n$
  - Case 1.1.2 a and b are both flips.
    We get that this is equivalent to counting pairs $(a_1, a_2), (b_1, b_2) \in \mathbb{Z}_n^2$ where $a_1 - b_1 \equiv x \mod n$ and $a_2 - b_2 \equiv y \mod n$
- Case 1.2, $a + b = (x, y, 0)$
  - Case 1.2.1 a and b are not flips.
    We get that this is equivalent to counting pairs $(a_1, a_2), (b_1, b_2) \in \mathbb{Z}_n^2$ where $a_1 + b_1 \equiv x \mod n$ and $a_2 + b_2 \equiv y \mod n$
  - Case 1.2.2 a and b are both flips.
    We get that this is equivalent to counting pairs $(a_1, a_2), (b_1, b_2) \in \mathbb{Z}_n^2$ where $a_1 - b_1 \equiv x \mod n$ and $a_2 - b_2 \equiv y \mod n$

Case 2, $F = 1$

- Case 2.1 $a - b = (x, y, 1)$
  - Case 2.1.1 a is a flip.
    We get that this is equivalent to counting pairs $(a_1, a_2), (b_1, b_2) \in \mathbb{Z}_n^2$ where $a_1 + b_1 \equiv x \mod n$ and $a_2 + b_2 \equiv y \mod n$
  - Case 2.1.2 b is a flip.
    We get that this is equivalent to counting pairs $(a_1, a_2), (b_1, b_2) \in \mathbb{Z}_n^2$ where $a_1 + b_1 \equiv x \mod n$ and $a_2 + b_2 \equiv y \mod n$
- Case 2.2 $a + b = (x, y, 1)$
  - Case 2.2.1 a is a flip.

We get that this is equivalent to counting pairs $(a_1, a_2), (b_1, b_2) \in \mathbb{Z}_n^2$ where $a_1 - b_1 \equiv x \mod n$ and $a_2 - b_2 \equiv y \mod n$

- Case 2.2.2 b is a flip.

  We get that this is equivalent to counting pairs $(a_1, a_2), (b_1, b_2) \in \mathbb{Z}_n^2$ where $a_1 + b_1 \equiv x \mod n$ and $a_2 + b_2 \equiv y \mod n$

We get a similar result with $a, b \in \mathbb{Z}_2 \ltimes \mathbb{Z}_{n^2}$ where $a - b = (xn + y, F)$ or $a + b = (xn + y, F)$ as follows:

Case 1, $F = 0$

- Case 1.1, $a - b = (xn + y, 0)$
  - Case 1.1.1 a and b are not flips.

    We get that this is equivalent to counting pairs $a_1n + a_2, b_1n + b_2 \in \mathbb{Z}_{n^2}$ where $(a_1n + a_2) - (b_1n + b_2) \equiv xn + y \mod n^2$
  - Case 1.1.2 a and b are both flips.

    We get that this is equivalent to counting pairs $a_1n + a_2, b_1n + b_2 \in \mathbb{Z}_{n^2}$ where $(a_1n + a_2) - (b_1n + b_2) \equiv xn + y \mod n^2$
- Case 1.2, $a + b = (xn + y, 0)$
  - Case 1.2.1 a and b are not flips.

    We get that this is equivalent to counting pairs $a_1n + a_2, b_1n + b_2 \in \mathbb{Z}_{n^2}$ where $(a_1n + a_2) + (b_1n + b_2) \equiv xn + y \mod n^2$
  - Case 1.2.2 a and b are both flips.

    We get that this is equivalent to counting pairs $a_1n + a_2, b_1n + b_2 \in \mathbb{Z}_{n^2}$ where $(a_1n + a_2) - (b_1n + b_2) \equiv xn + y \mod n^2$

Case 2, $F = 1$

- Case 2.1 $a - b = (xn + y, 1)$
  - Case 2.1.1 a is a flip.

    We get that this is equivalent to counting pairs $a_1n + a_2, b_1n + b_2 \in \mathbb{Z}_{n^2}$ where $(a_1n + a_2) + (b_1n + b_2) \equiv xn + y \mod n^2$
  - Case 2.1.2 b is a flip.

    We get that this is equivalent to counting pairs $a_1n + a_2, b_1n + b_2 \in \mathbb{Z}_{n^2}$ where $(a_1n + a_2) + (b_1n + b_2) \equiv xn + y \mod n^2$
- Case 2.2 $a + b = (xn + y, 1)$
  - Case 2.2.1 a is a flip.

    We get that this is equivalent to counting pairs $a_1n + a_2, b_1n + b_2 \in \mathbb{Z}_{n^2}$ where $(a_1n + a_2) - (b_1n + b_2) \equiv xn + y \mod n^2$
  - Case 2.2.2 b is a flip.

    We get that this is equivalent to counting pairs $a_1n + a_2, b_1n + b_2 \in \mathbb{Z}_{n^2}$ where $(a_1n + a_2) + (b_1n + b_2) \equiv xn + y \mod n^2$

8. S.J. Miller, B. Orosz, D. Scheinerman, Explicit constructions of infinite families of MSTD sets. J. Number Theory **130**(5), 1221–1233 (2010)
9. M.B. Nathanson, Problems in additive number theory. I, in *Additive Combinatorics*. CRM Proceedings and Lecture Notes, vol. 43 (American Mathematical Society, Providence, 2007), pp. 263–270
10. M.B. Nathanson, Sets with more sums than differences. Integers **7**(A5), 24 (2007)
11. Y. Zhao, Constructing MSTD sets using bidirectional ballot sequences. J. Number Theory **130**(5), 1212–1220 (2010)
12. Y. Zhao, Counting MSTD sets in finite abelian groups. J. Number Theory **130**(10), 2308–2322 (2010)
13. Y. Zhao, Sets characterized by missing sums and differences. J. Number Theory **131**(11), 2107–2134 (2011)

Geodetic Ratio, Color Sequences, and Their Relation on Graphs

Mustafa Atici, Matthew Hendrix, and Aathman Bhavaraju

Abstract Given a graph G, definitions of a color sequence and a (t, k)-geodetic set are presented in Atici (Graph Coloring: Color sequences and algorithm for color sequence, in *Proceedings of the 49-th Annual Association for Computing Machinery Southeast Conference, March 24–26 Kennesaw-GA* (2011), pp. 150–154) and (Ars Combin. **82**, 201–210 (2007)), respectively. In this paper, we define a new graph-theoretic parameter called *geodetic ratio*, which is derived from the definition of a (t, k)-geodetic. We show the relationship between the geodetic ratio and a color sequence, and give an algorithm to find an upper bound of the geodetic ratio of a given graph.

Keywords Geodesic · Geodetic set · Geodetic number · Geodetic ratio · Color sequence

MSC Codes: 05C85

1 Introduction

Let G be a simple undirected graph. The distance $d(u, v)$ between two vertices u and v in G is the length of the shortest $u - v$ path, the minimum number of edges between u and v. A $u - v$ path of length $d(u, v)$ is called a $u - v$ *geodesic*. Let $H(u, v)$ be the set of all vertices lying on a $u - v$ geodesic of G. The set $H(u, v)$ is called the *geodetic closure* of the $u - v$ geodesic.

M. Atici (✉)
Computer Science, School of Engineering and Applied Science, Western Kentucky University, Bowling Green, KY, USA
e-mail: mustafa.atici@wku.edu

M. Hendrix · A. Bhavaraju
The Gatton Academy of Mathematics and Science, Western Kentucky University, Bowling Green, KY, USA
e-mail: matthew.hendrix058@topper.wku.edu; aathmanbhavaraju874@topper.wku.edu

A. Bahmanian (ed.), *Combinatorics and Combinatorial Computing*, Springer Proceedings in Mathematics & Statistics 516,
https://doi.org/10.1007/978-3-032-08549-8_2

We illustrate $1-4$ geodesic and its closure $H(1,4)$ in the graph $G=(V,E)$, where $V(G)$={1, 2, 3, 4, 5} and $E(G)$={(1, 2), (2, 3), (3, 4), (4, 5), (5, 1), (1, 3)}. There are two $1-4$ geodesics since $d(1,4)=2$. These are $1-3-4$ and $1-5-4$. Hence, $H(1,4)=\{1,3,4,5\}$.

For any subset, $S \subseteq V(G)$, the geodetic closure of the set S is defined as follows.

$$H(S)=\bigcup_{u,v\in S} H(u,v).$$

A subset $S \subseteq V(G)$ is said to be a *geodetic set* of G if $H(S)=V(G)$, and the *geodetic number* $g(G)$ of graph G is defined as follows in [12]

$$g(G)=min\{|S| : \text{ S is a geodetic set of G }\}.$$

For example, the geodetic number of the above graph is $g(G)=3$. For more details about a geodetic set and the geodetic number of a graph, the interested reader may refer to [1, 3, 4, 10–14, 17].

For non-trivial graphs, any geodetic set with minimum size has at least two vertices. For example, a path with n vertices, P_n, has a geodetic set of size 2. On the other hand, a geodetic set with the largest size can have size n, where n is the number of vertices of the given graph G. For example, a complete graph on n vertices, K_n, has a geodetic set of size n. It was shown in [3] that the geodetic set problem is NP-complete.

The following definition stated in [4]. It is extended from the geodetic number, $g(G)$, of a graph G.

Definition 1 Let set S be a geodetic set of G with $|S|=k$ and $m(\langle S\rangle)=t$, where $\langle S\rangle$ is *vertex induced* subgraph of G and $m(\langle S\rangle)$ is the order of maximal component of graph $\langle S\rangle$. Then S is called a $(t,k)-$*geodetic set* of the graph G.

The rest of the paper is organized as follows: Next section, Sect. 2, defines the geodetic ratio and the geodetic ratio of some well-known graphs. Section 3 gives definitions of color sequence and the relation between geodetic ratio and color sequence. Section 4 gives a new upper bound for integrity, $I(G)$, of a graph G by using the new parameter geodetic ratio. In Sect. 5, we give a polynomial algorithm to determine an upper bound for the geodetic ratio ($gr(G)$) of any given simple and connected graph G. Note that in this paper, all the graphs are simple unless otherwise stated.

2 Geodetic Ratio

In graph theory, sometimes new definitions are derived from existing definitions to obtain better information about the topology of a given graph G. For example, *vertex-connectivity* (sometimes referred *connectivity*) of a graph G is the minimum cardinality of a set $U \subseteq V(G)$ such that $G-U$ has at least one more component

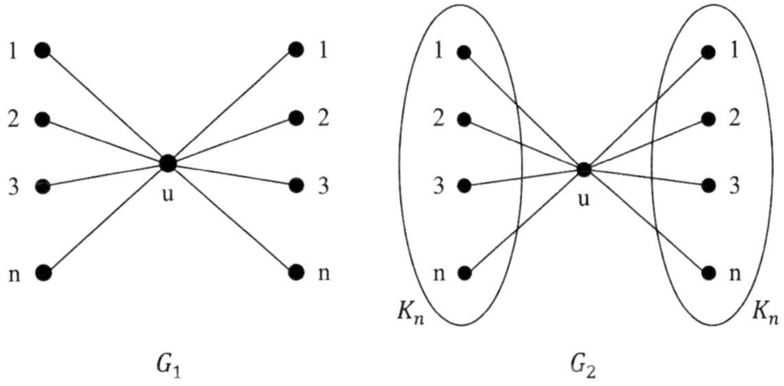

Fig. 1 Graphs G_1 and G_2 with same $g(G)$ but different $gr(G)$ numbers

than the graph G. It was pointed out in [9] that the connectivity parameter was not sufficient to explain the topologies of two different graphs even if they have the same connectivity numbers. Therefore, the definition of *integrity* was given. In this section, we define a new graph-theoretic parameter, the *geodetic ratio*, for a given graph G.

Definition 2 Let G be a graph. The *geodetic ratio* of G is defined as follows:

$$gr(G) = min_{S \subseteq V(G)}\{\frac{t}{k} : \text{ S is a } (t,k) - \text{geodetic set of G}\}.$$

Here are two examples to illustrate why such a new parameter definition is given.

Example 1 The following two graphs G_1, G_2, depicted in Fig. 1, are given. The graph G_1 is the star graph $K_{1,2n}$, and the graph G_2 is defined as follows: Take two copies of the complete graph K_n with a vertex u. Join the vertex u to every vertex of each K_n. Let S_i be a minimum geodetic set of G_i for $i = 1, 2$. Graph G_1 is a tree. All the end-vertices of the graph G_1 form its unique geodetic set S_1 [12]. So, we have $|S_1| = 2n$. Another theorem in [12] indicates that an upper bound for geodetic set size is $|S_2| \leq 2n + 1 - d + 1 = 2n$, where $d = 2$ is the diameter of G_2. The geodetic set S_2 must contain all the vertices in both K_n because they cannot be on any geodesic of length two. Therefore, $|S_2| = 2n$. S_1 is $(1, 2n)-$geodetic since $m(\langle S_1 \rangle) = 1$ and S_2 is $(n, 2n)-$ geodetic since $m(\langle S_2 \rangle) = n$. Therefore, these two graphs have $gr(G_1) = \frac{1}{2n}$ and $gr(G_2) = \frac{n}{2n} = \frac{1}{2}$, respectively.

The graphs G_1 and G_2 have the same geodetic numbers but different geodetic ratios. The first graph G_1 is very vulnerable to disruption. If the center vertex u is disabled, then the whole graph is separated into $2n$ components. None of these vertices can communicate with each other anymore. The second graph G_2, on the other hand, is not as vulnerable as the first one after removing the vertex u. The remaining n vertices in each component can communicate with each other.

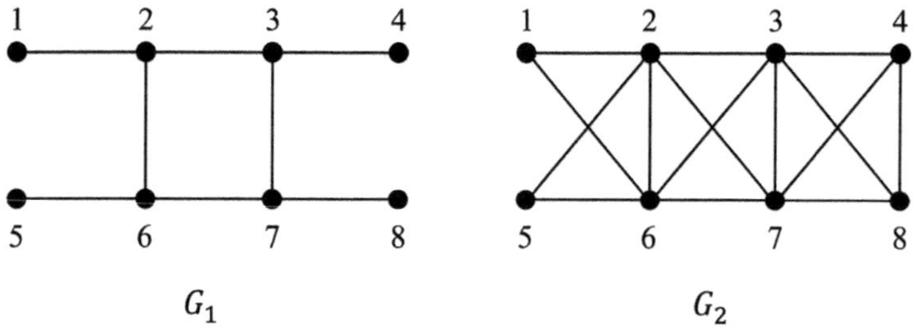

Fig. 2 Graphs G_1 and G_2 with same $g(G)$ but different $gr(G)$ numbers

Example 2 The following two graphs G_1, G_2, depicted in Fig. 2, are given.

Graphs G_1, G_2 have the same geodetic number, $g(G_1) = g(G_2) = 4$, but different geodetic ratio numbers, $gr(G_1) = \frac{1}{4}$ and $gr(G_2) = \frac{1}{2}$. This means that G_1 is more vulnerable to disruption than G_2. For example, if one can remove two vertices, 2 and 7, in graph G_1, then remaining graph has four components, with a maximum vertex of two in two components. However, if we remove any two vertices from G_2, the remaining graph is either connected or will have a component with at least four vertices.

One can observe from the above examples that the geodetic ratio, *gr(G)*, reviles more information than the geodetic number, $g(G)$, about the topology of given graph G. If $gr(G) = \frac{1}{k}$, where n is number of vertices of G and $k = \Theta(n)$, then the graph G is very vulnerable to disruption. That is, if one can remove small number of vertices, $G - S$, from the graph G, then the remaining graph consists of almost all single vertices. On the other hand, if $gr(G)$ is close to 1, then the graph is less vulnerable to disruption, as seen in the above two examples.

Using theorems given in [4], we can give the following theorem.

Theorem 1 *Let P_n be a path on n vertices. Then*

$$gr(P_n) = \begin{cases} \frac{2}{n+1} & \text{if } n \text{ is odd} \\ \\ \frac{2}{n} & \text{if } n \text{ is even} \end{cases} \tag{1}$$

Theorem 2 *Let T be a tree with n vertices. Then, the path P_n has a maximum $gr(G)$ among all the trees with n vertices. That is $gr(T) \leq gr(P_n)$.*

Proof Suppose vertex number n is even. Let T be a tree with n vertices. If T has $\frac{n}{2}$ or more end-vertex, then there is nothing to prove. Assume that S is a geodetic set of T with maximum size and $k < \frac{n}{2}$ is a number of all end-vertices in S. To maximize the number of vertices not in the geodetic set S, we assume that all the end-vertices in T have a distinct neighbor. These k neighbor vertices cannot be in S. There must be a vertex, say v, that is not an end-vertex and a neighbor of any of these k end-vertices since $2k < n$. Therefore, $S' = S \cup \{v\}$ is a new geodetic set with $|S'| = |S| + 1$. Contradiction to the fact that the geodetic set S has a maximum size. We can give a similar argument for an odd vertex number n. □

The geodetic ratio of the cycle C_n should be analyzed based on the number of vertices $n \geq 3$.

Theorem 3 *Let C_n be a cycle on $n \geq 3$ vertices. Then*

$$gr(C_n) = \begin{cases} 1 & \text{if } n \text{ is } 3 \\ \frac{2}{3} & \text{if } n \text{ is } 5 \\ \frac{2}{n-1} & \text{if} n \geq 7 \text{and odd} \\ \frac{2}{n} & \text{if} n \geq 4 \text{and even} \end{cases} \tag{2}$$

Proof Let us define vertex set to be $V(C_n) = \{v_1, v_2, \ldots, v_n\}$ and edge set to be $E(C_n) = \{(v_i, v_{i+1}) : i = 1, 2, \ldots, n-1\} \cup \{(v_1, v_n)\}$.
Case 1: If $n = 2k \geq 4$, then let $S = \{v_1, v_3, \ldots, v_{n-1}\}$, a set of all vertices with an odd index. Set S is a geodetic set of C_n since any vertex in $V(C_n) - S$ is on a geodesic of length two in S. $|S|$ is maximum with $m(\langle S \rangle) = 1$ since adding one more vertex to the set S will create a component with order at least two in $\langle S \rangle$. Therefore, $gr(C_n) = \frac{2}{n}$.
Case 2: If $n = 2k + 1 \geq 7$, then let $S = \{v_2, v_4, \ldots, v_{2k}\}$, a set of all vertices with even index. This is a geodetic set of C_n with $|S| = \frac{n-1}{2}$. Vertices v_1 and v_{2k+1} are on $(v_2, v_{2k})-$geodesic since $d(v_2, v_{2k}) = 3$. On the other hand, if we add any vertex with odd indexes to the set S, then $\langle S \rangle$ will have a component with a vertex number of at least 2. Hence, $gr(C_n) = \frac{2}{n-1}$.
Case 3: If $n = 3$, then the cycle graph is K_3. Therefore, the geodetic ratio is 1.
Case 4: If $n = 5$, then set $S = \{v_2, v_4\}$ is not a minimum geodetic set since v_1 and v_5 cannot be on any geodesic of length two. Either v_1 or v_5 must be in the set S for it to be a geodetic set. Therefore, the geodetic ratio of C_5 is $\frac{2}{3}$. □

Once we have determined the geodetic ratio of a cycle C_n, we easily derive the following conclusion:

Corollary 1 *Let G be a Hamiltonian graph on n vertices. Then*

$$gr(G) \geq gr(C_n) \tag{3}$$

Proof We assume that vertex number n in G is even. Let S be a $(t, k)-$geodetic set of G so that $gr(G) = \frac{t}{k}$. There exists a cycle C containing all n vertices of G. If $t = 1$, then $k \leq \frac{n}{2}$ because the set S cannot have more than half of the vertices of the cycle C. If $t = 2$, then $\frac{2}{k} < \frac{2}{n}$ if and only if $k > n$. We can give a similar argument for an odd number of the vertices. □

Using the argument in the geodetic ratio of cycle C_n, we give the following result for wheel graph W_n.

Theorem 4 *Let W_n be a wheel graph on n vertices. Then*

$$gr(W_n) = \begin{cases} \frac{2}{n-1} \text{ } if n \geq 5 and\ odd \\ \\ \frac{4}{n} \text{ } if n \geq 4 and\ even \end{cases} \tag{4}$$

Proof The wheel graph W_n has cycle C_{n-1} and a "center" vertex, say v, which connects to all the vertices on the cycle C_{n-1}. If $n \geq 5$ is odd, then $gr(W_n) = gr(C_{n-1})$. If $n \geq 6$ is even, then a geodetic set S must have all odd indexed vertices in C_{n-1} since the distance between any two vertices in W_n is two. Therefore, S is a geodetic set with $|S| = \frac{n}{2}$ vertices with $m(\langle S \rangle) = 2$. In particular for $n = 4$, $W_4 = K_4$. Given formula $\frac{4}{4} = 1$ still holds. □

The following theorem is for a generalized complete binary tree. It is a full k-ary tree T_n^k with $\frac{k^{n+1}-1}{k-1}$ vertices and height n. In a full k-ary tree, every internal vertices have k children.

Theorem 5 *Let T_n^k be a full k-ary tree. Then*

$$gr(T_n^k) = \begin{cases} \frac{k^2-1}{k^{n+2}-1} \text{ } if\ n\ is\ even \\ \\ \frac{k^2-1}{k(k^{n+1}-1)} \text{ } if\ n\ is\ odd \end{cases} \tag{5}$$

Proof Let us first redefine the set of vertices of T^k as follows:

1. Set S_0 contains the root vertex with $|S_0| = 1$.
2. Set S_i contains all the vertices that have distance i from the root with $|S_i| = k^i$.

Observations on S_i:

1. There is no edge between vertices in S_i for $i = 1, 2, \ldots, n$.
2. There is no edge between any vertex in set S_i to any vertex in S_{i+k}, where $k \geq 2$ for $i = 0, 1, \ldots, n-2$.
3. An edge exists only between a vertex in S_i to a vertex in S_{i+1} for $i = 0, 1, 2, \ldots, n-1$.
4. if $n = 2r$ for some integer r, then $\sum_{i=0}^{r} |S_{2i}| = 1 + k^2 + k^4 + k^6 + \ldots + k^{2r} = \sum_{i=0}^{r} (k^2)^i = \frac{(k^2)^{r+1}-1}{k^2-1} = \frac{k^{n+2}-1}{k^2-1}$
5. if $n = 2r+1$ for some integer r, then $\sum_{i=1}^{r} |S_{2i+1}| = k + k^3 + k^5 + k^7 + \ldots + k^{2r+1} = k \times \sum_{i=0}^{r} (k^2)^i = k \times \frac{(k^2)^{r+1}-1}{k^2-1} = k \times \frac{k^{(2r+1)+1}-1}{k^2-1} = k \times \frac{k^{n+1}-1}{k^2-1}$

We can think of the given graph as a *path* on vertices S_i for $i = 0, 1, 2, \ldots, n$. Similar to the proof of $gr(P_n)$; if n is even, then we can show that set $S = S_0 \cup S_2 \cup \ldots \cup S_n$ is geodetic set with $|S| = \frac{k^{n+2}-1}{k^2-1}$ and $m(\langle S \rangle) = 1$. Therefore, $gr(T_n^k) \leq$

$\frac{k^2-1}{k^{n+2}-1}$. Assume S' is another geodetic set with $|S| < |S'|$. The geodetic set S' must have at least two vertices, say x, y, where if $x \in S_i$ for some i, then $y \in S_{i+1}$. This yields $m(\langle S' \rangle) > k$; and $\frac{k^2-1}{k^{n+2}-1} \leq \frac{k}{|T_n^k|} = \frac{k(k-1)}{k^{n+1}-1}$ if $k \geq 2$.
We can give a similar argument can be given if n is odd. Then the geodetic set $S = S_1 \cup S_3 \cup \ldots \cup S_n$ with $|S| = k \times \frac{k^{n+1}-1}{k^2-1}$ and $m(\langle S \rangle) = 1$. □

A binomial tree B_k is a tree with 2^k vertices such that B_k has height k and the number of vertices in each level is $\binom{k}{i}$ for all $i = 0, 1, \ldots, k$. By induction, we can easily show that B_k has 2^{k-1} leaf nodes. Hence, we have the following theorem.

Theorem 6 *Let B_k be a binomial tree. Then*

$$gr(B_k) = \frac{1}{2^{k-1}}. \tag{6}$$

If a bipartite graph $G = (V, E)$ is not connected, then giving an upper bound for geodetic ratio $gr(G)$ is not possible in terms of the cardinality of V_1 or V_2, where $V(G) = V_1 \cup V_2$. If the bipartite graph G is connected, we can have the following upper bound on the geodetic ratio.

Theorem 7 *Let G be a connected bipartite graph with $V(G) = V_1 \cup V_2$, where $|V_1| \geq |V_2|$. Then $gr(G) \leq \frac{1}{|V_1|}$.*

Proof We have two cases to look at.
Case 1: All vertices in V_2 have a degree of at least two. If we take $S = V_1$, then S is a geodetic set, and $gr(G) = \frac{1}{|V_1|}$.
Case 2: There exists at least one vertex in V_2 with degree one. Define $U = \{u \in V_2 : deg_G(u) = 1\}$ and $N(U)$ to be all adjacent vertices of $U \subseteq V_2$. Since G is connected, $|U| \geq |N(U)|$ and $V_1 - N(U) \neq \phi$. Now, define set S as follows: $S = U \cup V_1 - N(U)$. We will show that S is a geodetic set with $m\langle S \rangle = 1$. There are two groups of vertices that we need to show to be on a geodesic. These are vertices in $V_2 - U$ and $N(U)$. First, let $v_2 \in V_2 - U$. We have the following cases:

1. We know that $deg(v_2) \geq 2$. If at least two neighbors are in $V_1 - N(U)$, then v_2 is on a geodesic of two vertices.
2. If v_2 has a single neighbor in $V_1 - N(U)$, say w_1, and all others in $N(U)$. Assume one of the neighbors in $N(U)$ is u_1. There is a vertex in U, say u_2, so that $(u_1, u_2) \in E(G)$. Hence, v_2 is on the $w_1 - u_2$ geodesic.
3. If all neighbors of v_2 are in $N(U)$. Assume two of these neighbors are u_1, u_1'. There are two vertices in U, say u_2, u_2', so that (u_1, u_2) and (u_1', u_2') are in $E(G)$. Hence, v_2 is on $u_1' - u_2'$ geodesic.

Next, let $v_1 \in N(U)$. We have the following cases:

1. If v_1 has two neighbors in U, then v_1 is on a geodesic of two vertices in U.

2. If v_1 has a unique neighbor in U, say u_1. We know that $V_1 - N(U) \neq \phi$. Assume that $x \in V_1 - N(U)$. Since G is connected and $deg_G(u_1) = 1$, v_1 must be on a $u_1 - x$ geodesic.

S is a geodetic set with $m\langle S\rangle = 1$ and $|S| = |V_1| + |U| - |N(U)| \geq |V_1|$ □

Let t and m be any two positive integers with $1 \leq t \leq m$. We now show the existence of a graph G with $gr(G) = \frac{t}{m}$.

Theorem 8 *For any given positive integers m and $1 \leq t \leq m$, there exists a simple graph G with $gr(G) = \frac{t}{m}$.*

Proof We have two cases to consider:

Case 1: if $t = m$, then take the complete graph $G = K_m$ which has $gr(G) = \frac{m}{m} = 1$.

Case 2: if $t < m$, let K_t be the complete graph on t vertices. Pick $k + 1$ vertices $\{u, v_1, v_2, \ldots, v_k\}$, where $k + t = m$. Construct the graph G as follows: Vertex set of G is $V(G) = V(K_t) \cup \{u, v_1, v_2, \ldots, v_k\}$ with $|V(G)| = m+1$. The edge set of G is $E(G) = E(K_t) \cup \{(w, u)|$ for every vertex w of $K_t\} \cup \{(u, v_i)$: for $i = 1, 2, \ldots, k\}$ with $|E(G)| = \frac{t(t+1)}{2} + k$. If two vertices x, y, are not adjacent in the graph G, then the distance between x, y, is $d(x, y) = 2$. Since all vertex v_i's have degree 1, they must be in a geodetic set S. Also, none of the vertices of K_t can be on any path of length 2. Hence, $S = V(K_t) \cup \{v_1, v_2, \ldots, v_k\}$ is a geodetic set with minimum $|S| = t + k = m$ and $gr(G) = \frac{t}{m}$. □

3 Geodetic Ratio and Maximum Color Sequence

The chromatic number of a graph G is the smallest number of colors needed to color the vertices of G so that two adjacent vertices cannot have the same color. The notation of the chromatic number of a graph G is $\chi(G)$. Determining the chromatic number of a graph is a well-known and much-studied problem in graph theory. By using vertex coloring, we summarize definitions of color sequence and maximum color sequence of a given graph G, as stated in [5].

Definition 3 Let G be a graph colorable with k colors. The finite sequence $cs(G) = (n_1, n_2, \ldots, n_k)$ is a *color sequence* of a given graph G in which $n_1 \geq n_2 \geq, \ldots, \geq n_k$, where color i used n_i times to color the vertices of G for $i = 1, 2, 3, \ldots, k$.

Let us define the lexicographical order of two color sequences before the maximum color sequence definition. We say color sequence $cs(G) = (n_1, n_2, \ldots, n_k)$ is *lexicographically bigger* than the color sequence $cs(G) = (m_1, m_2, \ldots, m_t)$ if there exists an integer j, where $j \leq min\{k, t\}$, such that $n_i = m_i$ for all $i = 1, 2, \ldots, j - 1$ and $n_j > m_j$.

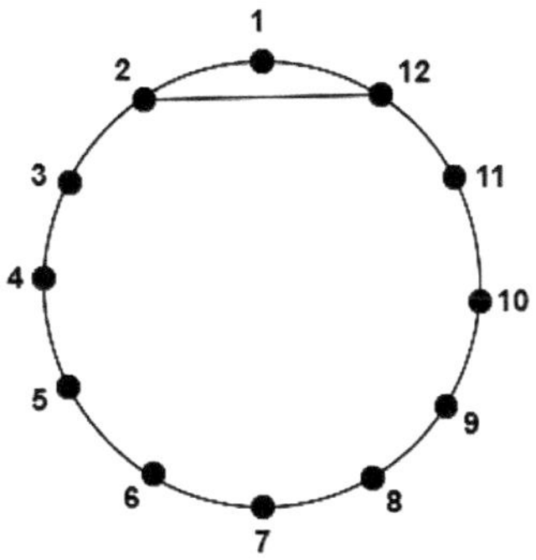

Fig. 3 Graph with six color sequences of length three

Definition 4 Let G be a graph colorable with k colors. If $cs(G)=(m_1, m_2, \ldots, m_k)$ is the maximum of all the color sequences of the given graph G in lexicographical order, then it is called *maximum color sequence* and denoted by $msc(G) = (m_1, m_2, \ldots, m_k)$.

By the definition of color sequence, it would be more meaningful to talk about maximum color sequence among all the color sequences that have the same length.

Here are some examples of color sequence and maximum color sequence:

1. Star graph $K_{1,n}$ has $mcs(K_{1,n}) = (n, 1)$.
2. The path P_n on n vertices $mcs(P_n) = (\frac{n}{2}, \frac{n}{2})$ if n is even and $mcs(P_n) = (\frac{n+1}{2}, \frac{n-1}{2})$ if n is odd.
3. The cycle C_n on n vertices $mcs(C_n) = (\frac{n}{2}, \frac{n}{2})$ if n is even and $mcs(C_n) = (\frac{n-1}{2}, \frac{n-1}{2}, 1)$ if n is odd.
4. The complete graph on n vertices $mcs(K_n) = (1, 1, \ldots, 1)$.
5. The following graph G, depicted in Fig. 3, has six different color sequence of length three. These color sequences are: $cs_1(G) = (4, 4, 4)$, $cs_2(G) = (5, 4, 3)$, $cs_3(G) = (5, 5, 2)$, $cs_4(G) = (6, 5, 1)$, $cs_5(G) = (6, 3, 3)$, and $cs_6(G) = (6, 4, 2)$. Among those, the maximum color sequence is $mcs(G) = (6, 5, 1)$.

The following theorem is about the existence of a graph for any given color sequence. For any given integers $n_1, n_2, \ldots, n_k$ we can construct complete $k-$ partite graph.

Theorem 9 *Let $(n_1, n_2, \ldots, n_k)$ be a given color sequence. Then there is a graph G with $mcs(G) = (n_1, n_2, \ldots, n_k)$.*

Now, we turn our attention to the relation between the maximum color sequence and geodetic ratio of a given graph G.

Theorem 10 *Let G be a connected bipartite graph with $mcs(G) = (n_1, n_2)$. Then $gr(G) = \frac{1}{n_1}$.*

5 Algorithm to Compute an Upper Bound of $gr(G)$

Computing the geodetic number of a graph is an NP-complete problem [3]. Therefore, finding a polynomial algorithm to determine the geodetic ratio of a given graph is not possible at this point. We present an algorithm to find an upper bound for $gr(G)$ when a simple and connected graph G.

The algorithm uses a well-known *Breadth-first search(BFS)* algorithm in graph theory. Given a graph $G = (V, E)$ and a *source* vertex $s \in V(G)$, breadth-first search explores the edges of G to visit every vertex that is reachable from s. It computes the distance from s to each reachable vertex. It also produces a *breadth-first tree(BFT)* with root s, the path in the breadth-first tree from s to v corresponds to $d(s, v)$ in G.

For a given connected graph G and its BFT tree T with source s and height $k \geq 1$, we regroup vertices as follows: $A_0 = \{s\}$ and $A_i = \{v : d(s, v) = i \text{ in } G\}$. Further, $B_i = \{v : v \in A_i \text{ and } deg_G(v) = 1 \text{ or } v \text{ has no adjacent vertices in } A_{i+1}\}$ for $i = 1, 2, \ldots, k-1$.

Lemma 1 *Let G be a connected graph and T be its BFT with height $k > 0$. Then $S = A_0 \cup A_k \cup_{i=1}^{k-1} B_i$ is a geodetic set of G.*

Proof Let $v_i \in A_i$, where $1 \leq i \leq k-1$. If $v_i \notin B_i$, then v_i is on a $s - v$ geodesic, where $v \in A_k$. If $v_i \in B_i$, then it is on a $s - v_i$. □

The geodetic ratio(gr), which obtained from the set S in Lemma 1, can be improved for some graph G. By the construction of BFT tree T, there is no edge between any vertices in A_i and A_{i+2} in G for $i = 0, 1, \ldots, k-2$. Therefore, we may be able to add alternating A_i's to the starting geodetic set S as long as $\frac{m(\langle S \cup A_i \rangle)}{|S \cup A_i|} \leq \frac{m(\langle S \rangle)}{|S|}$.

The following algorithm, *Geodetic Ratio Computation(GRC)*, receives a connected graph G and returns an upper bound for $gr(G)$. Geodetic Ratio Computation(GRC) is a greedy algorithm.

GRC(G)

$gr = 1$
for each vertex $s \in V(G)$
    Run *BFS(s,G)* to form *BFT T*
    Construct sets $A_0, A_1, \ldots, A_k, B_1, B_2, \ldots, B_{k-1}$
    Initialize $S = A_0 \cup A_k \cup_{i=1}^{k-1} B_i$ and $gr(G) = \frac{m(\langle S \rangle)}{|S|}$
    /*check height k of T and determine which A_i's can be added to S*/
    if $k = 1$ /*$S = A_0 \cup A_1$*/
        if $u, v \in A_1$ with $d(u, v) = 2$
            then $S = A_1$ and update $gr(G)$
    else if $k \geq 3$ and k is odd
        for $j = r - 1$ down to 3 /*$k = 2r + 1$ for an integer r*/
            if $\frac{m(\langle S \cup A_{2j-1} \rangle)}{|S \cup A_{2j-1}|} \leq \frac{m(\langle S \rangle)}{|S|}$
                then $S = S \cup A_{2j-1}$ and update $gr(G)$

Fig. 5 Graph G and its BFT tree T

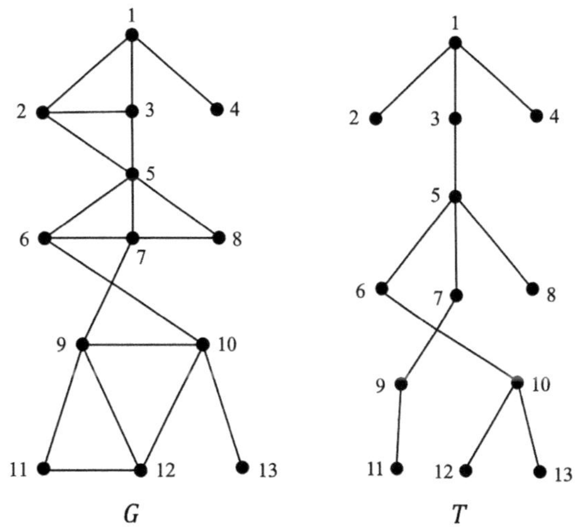

/*after loop is over, we can check if A_1, A_0 can be swapped in S*/

13 **if** $u, v \in A_1$ with $d(u, v) = 2$ and $\frac{m(\langle S \cup A_1 - A_0 \rangle)}{|S \cup A_1| - 1} \leq \frac{m(\langle S \rangle)}{|S|}$

14 **then** $S = S \cup A_1 - A_0$ and update $gr(G)$

15 **else** /*$k = 2r$ for an integer r*/

16 **for** $j = 1$ to $r - 1$

17 **if** $\frac{m(\langle S \cup A_{2j} \rangle)}{|S \cup A_{2j}|} \leq \frac{m(\langle S \rangle)}{|S|}$

17 **then** $S = S \cup A_{2j}$ and update $gr(G)$

18 **if** $gr(G) < gr$

19 **then** $gr = gr(G)$

20 **return** gr

Example 5 In the following graph G, we illustrate how the algorithm computes an upper bound gr for the geodetic ratio $gr(G)$. Figure 5 shows graph G and its BFT tree T. Here are the set A_i's and B_i's: $A_0 = \{1\}$, $A_1 = \{2, 3, 4\}$, $A_2 = \{5\}$, $A_3 = \{6, 7, 8\}$, $A_4 = \{9, 10\}$, $A_5 = \{11, 12, 13\}$ and $B_1 = \{4\}$, $B_2 = \phi$, $B_3 = \{8\}$, $B_4 = \phi$. The vertex 4 is in B_1 but not the vertex 2 because vertex $3 \in A_2$ is a neighbor of vertex 2.

In the first iteration, when root vertex $s = 1$, the above tree T is constructed in line 3. In lines $4 - 5$, sets $A_0, A_1, \ldots, A5, B_1, \ldots, B_4$ are defined, and $S = A_0 \cup A_5 \cup B_1 \cup B_3 = \{1, 4, 8, 11, 12, 13\}$ is initialized. Since $m(\langle S \rangle) = 2$ and $|S| = 6$, we have $grG) = \frac{2}{6} = \frac{1}{3}$. The height of the tree T is $k = 5$. In line 9, we check if set $A_3 = \{6, 7, 8\}$ can be added to the set S. A_3 cannot be added to S since $\frac{m(\langle S \cup A_3 \rangle)}{|S \cup A_3|} = \frac{3}{8} \not\leq \frac{2}{6}$ Next, at line 13, we check if A_1 and A_0 can be swapped,i.e., add A_1 to S while removing A_0. A_1 has two vertices, 2 and 4 with $d(2, 4) = 2$ and $\frac{m(\langle S \cup A_1 - A_0 \rangle)}{|S \cup A_1 - 1|} = \frac{2}{7} < \frac{3}{6}$. Therefore, $S = S \cup A_1 - A_0 = \{2, 3, 4, 8, 11, 12, 13\}$ and $gr(G) = \frac{2}{7}$ at this point. At the first iteration in line 19 $gr = \frac{2}{7}$ is updated.

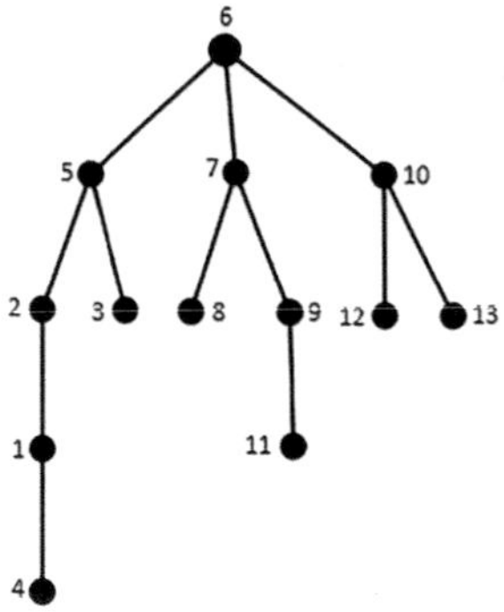

Fig. 6 *BFT* tree T with root $s = 6$ and height $k = 4$

Table 1 Geodetic ratio computation

Root	Current geodetic Set	Current geodetic ratio
1	$S = \{2, 3, 4, 8, 11, 12, 13\}$	$gr(G) = 2/7$
2	$S = \{2, 3, 4, 8, 11, 12, 13\}$	$gr(G) = 2/7$
3	$S = \{2, 3, 4, 8, 11, 12, 13\}$	$gr(G) = 2/7$
4	$S = \{2, 3, 4, 8, 11, 12, 13\}$	$gr(G) = 2/7$
5	$S = \{4, 5, 8, 11, 12, 13\}$	$gr(G) = 2/7$
6	$S = \{4, 6, 8, 11, 13\}$	$gr(G) = 1/5$
7	$S = \{4, 7, 8, 11, 12, 13\}$	$gr(G) = 1/5$
8	$S = \{4, 8, 11, 13\}$	$gr(G) = 1/5$
9	$S = \{4, 6, 8, 9, 11, 12, 13\}$	$gr(G) = 1/5$
10	$S = \{4, 6, 8, 10, 11, 13\}$	$gr(G) = 1/5$
11	$S = \{4, 6, 8, 11, 13\}$	$gr(G) = 1/5$
12	$S = \{2, 3, 4, 8, 11, 12, 13\}$	$gr(G) = 1/5$
13	$S = \{2, 3, 4, 8, 11, 12, 13\}$	$gr(G) = 1/5$

At the sixth iteration, *BFT* three T, in Fig. 6, with root $s = 6$ and height $k = 4$, is constructed. $A_0 = \{6\}$, $A_1 = \{5, 7, 10\}$, $A_2 = \{2, 3, 8, 9, 12, 13\}$, $A_3 = \{1, 11\}$, $A_4 = \{4\}$, $B_2 = \{8, 13\}$, and $B_3 = \{11\}$. Initial geodetic set $S = \{4, 6, 8, 11, 13\}$ with $gr(G) = \frac{1}{5}$. At line 14, we check if A_2 can be added to S to obtain a lower ratio. After the loop is over in line 1, the algorithm returns the geodetic ratio $gr = \frac{1}{5}$ for the given graph G. Indeed, this is the optimal geodetic ratio for this graph. Table 1 lists current root, geodetic set, and geodetic ratio for all thirteen iterations.

We determine the complexity of the above algorithm as follows. Let $G = (V, E)$ be a given graph with $|V| = V(G)$ vertices and $|E| = E(G)$ edges. Since G is assumed to be a simple graph, we have $|E| \leq |V|(|V| - 1)/2$. Here is an analysis of the main lines of the algorithm:

1. In line 3, complexity of $BFS(s, G)$ to construct the BFT tree T is $O(|V|+|E|)$.
2. A_i's can be constructed during the construction of the tree T.
3. We can form the set B_i's, after A_i's constructed, in no more than $O(|V| + |E|)$ steps.

4. Lines $6 - 8$ will not take more than $|V|((|V| - 1)/2$ steps.
5. In line 11, $m(\langle S \rangle)$ can be computed using the *depth-first search* algorithm which has $\Theta(|V| + |E|)$ complexity. We may repeat this line at most $k <= n/2$ times.
6. In line 15, we have similar complexity as in line 11.

The complexity of the above algorithm is $\Theta(|V|^3 + |V|^2|E|)$ since there is a loop in line 2. This is the first polynomial algorithm to compute upper bounds for $g(G)$, $I(G)$, and $gr(G)$ of any simple connected graph G.

References

1. M. Atici, Graph operations and geodetic numbers. Congressus Numer. **141**, 95–110 (1999)
2. M. Atici, Integrity of regular graphs and integrity graphs. J. Combin. Math. Combin. Comput. **37**, 27–42 (2001)
3. M. Atici, Computational complexity of geodetic set. Int. J. Comput. Math. **79**(5), 587–591 (2002)
4. M. Atici, (t,k)-Geodetic set of a graph. Ars Combin. **82**, 201–210 (2007)
5. M. Atici, Graph Coloring: Color sequences and algorithm for color sequence, in *Proceedings of the 49-th Annual Association for Computing Machinery Southeast Conference, March 24–26 Kennesaw-GA* (2011), pp. 150–154
6. M. Atici, R. Crawford, C. Ernst, The integrity of n-cage graphs. Int. J. Comput. Math. **81**(11), 1341–1348 (2004)
7. K.S. Bagga, L.W. Beineke, W.D. Goddard, M.J. Lipman, R.E. Pippert, A survey of integrity. Discrete Appl. Math. **37–38**, 13–28 (1992)
8. C.A. Barefoot, R. Entringer, H.C. Swart, ıIntegrity of trees and powers of cycles. Congr. Numer. **58**, 103–114 (1987)
9. C.A. Barefoot, R. Entringer, H.C. Swart, Vulnerability in graphs-A comparative survey. J. Combin. Math. Combin. Comput. **1**, 12–22 (1987)
10. S.R. Canoy, Jr., I.J.L. Garces, Convex sets under some graph operations. Graphs Combin. **18**, 787–793 (2002)
11. G. Chartrand, P. Zhang, Convex sets in graphs. Congr. Numer. **136**, 19–32 (1999)
12. G. Chartrand, F. Harary, P. Zhang, Geodetic sets in graphs. Discuss. Math. Graph Theory **20**(1), 129–138 (2000)
13. G. Chartrand, F. Harary, P. Zhang, On the geodetic number of a graphs. Networks **39**(1), 1–6 (2000)
14. G. Chartrand, C. Wall, P. Zhang, The convexity number of a graph. Graphs Combin. **18**, 209–217 (2002)
15. H.L. Clarc, R.C. Entringer, M.R. Fellows, Computational complexity of integrity. J. Combin. Math. Combin. Comput. **2**, 179–191 (1987)
16. W. Goddard, H.C. Swart, Integrity in graphs: bounds and basic. J. Combin. Math. Combin. Comput. **7**, 139–151 (1990)
17. F. Harary, E. Loukakis, C. Tsouros, The geodetic number of a graph. Math. Comput. Model. **17**(11), 89–93 (1993)
18. A. Vince, The integrity of a cubic graph. Discrete Appl. Math. **140**(1–3), 223–239 (2004)

people. Participants are requested to list the $s - p = 1$ hour when they would prefer to take a break (hours around lunch time or the first and last hours to arrive late or leave early being likely popular choices). This information is described by a vector $V = (c_1, c_2, \ldots, c_s)$, the ith component being the number of participants wanting to meet during the ith of the s hours (so $v - c_i$ is the number of participants wanting their break during the ith hour).

This problem can be modeled by using the complete graph K_v, each vertex representing a participant, with the edge $\{i, j\}$ being colored k to indicate that participants i and j should meet during the kth hour. To meet the requirements, at each vertex w, each of some choice of p colors should appear on exactly $(v-1)/p$ edges incident to w, the remaining $s - p$ colors appearing on no incident edges. Furthermore, to meet the choice of when to take a break, the number of vertices missing color k should be $v - c_k$ for $1 \leq k \leq s$ where c_k is the kth component of V. As will be defined below, this is an equitable (s, p)-edge-coloring of K_v, say E, with associated color vector $V(E)$.

Completing the scheduling would then require, for $1 \leq k \leq s$, taking the subgraph $G(k)$ induced by the edges colored k and giving it an edge-coloring with $\max\{(v-1)/p, \chi'(G(k))\}$ colors to subdivide the kth hour into meeting times. But this step is not the focus of the research in this paper.

Although this paper focuses on edge-colorings, historically, the problem is stated in terms of design theory. We will start with that history. An *H-decomposition* of a graph G is an ordered pair (V, B) where V is the vertex set of G and B is a partition of the edges of G into sets, each of which induces a copy of H. The graphs induced by the elements of B are known as the *blocks* of the decomposition. (V, B) is said to have an *(s, p)-block-coloring* $E : B \to C = \{1, 2, \ldots, s\}$ if:

1. the blocks in B are colored with exactly s colors, and
2. for each vertex $u \in V(G)$ the blocks containing u are colored using exactly p colors.

The (s,p)-block-coloring, E, is said to be *equitable* if

3. for each vertex $u \in V(G)$ and for each $\{i, j\} \subset C(E, u)$, $|b(E, u, i) - b(E, u, j)| \leq 1$,

where $C(E, u) = \{i \mid E \text{ colors some block incident with } u \text{ with color } i\}$, and $b(E, u, i)$ is the number of blocks in B containing u that are colored i by E. Note that this definition of equitable generalizes the definition used in the usual edge-coloring situation where the blocks are copies of K_2 and $s = p$.

Such colorings were first introduced by Colbourn and Rosa who considered a more general notion in regards to Steiner Triple Systems (K_3-decompositions of K_v) in [2]. M. Gionfriddo et al. were the first to consider the equitable block-colorings as defined above, studying equitable block-colorings of Steiner Triple Systems in [5].

L. Gionfriddo, M. Gionfriddo, and Ragusa considered the existence of equitable block-colorings of 4-cycle systems of K_v in [6]. M. Gionfriddo and Ragusa extended their work in [4]. In [9], Li and Rodger considered the existence of equitable (s, p)-block-colorings of C_4-decompositions of $K_v - F$, where F is a 1-factor of K_v, allowing them to consider even values of v previously excluded by necessary conditions. This work was extended by Li, Matson, and Rodger in [10] and by Matson and Rodger in [11]. We extend an area of their work in this paper.

For any C_4-decomposition $\Sigma = (V, B)$ of $K_v - F$, the authors in [9] defined the *spectrum* of Σ as $\Omega_p(\Sigma) = \{s \mid$ there exists an equitable (s, p)-block-coloring of $\Sigma\}$. They defined the *p-color-spectrum* $\Omega_p(v) = \cup\Omega_p(\Sigma)$, the union being taken over all C_4-decompositions, Σ, of $K_v - F$. Two values in $\Omega_p(v)$ are of particular interest: the *lower p-chromatic index* is defined to be $\chi'_p(v) = \min \Omega_p(v)$ and the *upper p-chromatic index* is defined to be $\overline{\chi'_p(v)} = \max \Omega_p(v)$. In this paper we will focus on a particularly interesting case where, as described in the following result, the lower p-chromatic index is not equal to p; i.e., $\chi'_p(v) > p$. It's interest is generated for edge-colorings (i.e., K_2-decompositions where it is also the case that $\chi'_p(v) > p$) by the observation that, because at each vertex a color is prohibited from appearing on an incident edge, the usual edge-coloring technique of interchanging colors along a 2-edge-colored path does not work. So completely new techniques are needed.

Theorem 1 ([9, 10]) *For any edge-coloring (K_2-decomposition) of K_v,*

$$\chi'_{2t}(v) = \begin{cases} 2t+1 & \textit{if } v \equiv 2t+1 \pmod{4t} \\ 2t & \textit{otherwise.} \end{cases}$$

For any C_4-decomposition of $K_{2v} - F$,

$$\chi'_{2t}(2v) = \begin{cases} 2t+1 & \textit{if } 2v \equiv 4t+2 \pmod{8t} \\ 2t & \textit{otherwise.} \end{cases}$$

We will examine this intriguing case more closely. Thus, for the rest of this paper we will assume $v \equiv 2t + 1 \pmod{4t}$, $p = 2t$, and $s = 2t + 1$.

In [10], the authors began to focus on the structure of these equitable (s, p)-block-colorings using the following concepts. The *color vector* of an equitable (s, p)-block-coloring E of an H-decomposition $(V(G), B)$ of a graph G is the vector $V(E) = (c_1(E), c_2(E), \ldots, c_s(E))$ in which, for $1 \leq i \leq s$, $c_i(E)$ is the number of vertices in G that are incident with a block of color i; by renaming colors if necessary, it is always assumed that $c_1(E) \leq c_2(E) \leq \cdots \leq c_s(E)$. The following definitions help us explore the color vector. For any graphs G and H and for $1 \leq i \leq s$, define

(i) $\phi(H, G; s, p, i) = \{c_i(E) \mid E$ is an equitable (s, p)-block-coloring of an H-decomposition of $G\}$,
(ii) $\underline{\psi'}(H, G; s, p, i) = \min \phi(H, G; s, p, i)$, and
(iii) $\overline{\psi'}(H, G; s, p, i) = \max \phi(H, G; s, p, i)$.

Because we are only considering $s = 2t + 1$ and $p = 2t$, we define $\psi_i'(H, G) = \psi'(H, G; 2t + 1, 2t, i)$ and $\overline{\psi_i'}(H, G) = \overline{\psi'}(H, G; 2t + 1, 2t, i)$.

The motivation behind this work stems from design theory. In [10], the values of $\overline{\psi_1'}(C_4, K_{2v} - F)$ and $\psi_{2t+1}'(C_4, K_{2v} - F)$ were found. In [11], $\psi_1'(C_4, K_{2v} - F)$ and $\overline{\psi_{2t+1}'}(C_4, K_{2v} - F)$ were found, along with $\overline{\psi_i'}(C_4, K_{2v} - F)$ for $2 \leq i \leq 2t$. In this paper we find $\psi_2'(C_4, K_{2v} - F)$ (see Theorem 10), $\psi_t'(C_4, K_{2v} - F)$ (see Theorem 12), and $\psi_{2t}'(C_4, K_{2v} - F)$ (see Theorem 9).

We use the methods established in [9–11] to find $\psi_i'(C_4, K_{2v} - F)$ by considering $\psi_i'(K_2, K_v)$. Thus, most of the paper is focused on equitable (s, p)-edge-colorings of K_v, finding $\psi_{2t}'(K_2, K_v)$ (see Theorem 5), $\psi_2'(K_2, K_v)$ (see Theorem 6), and $\psi_t'(K_2, K_v)$ (see Theorem 8). This focus on these three special values of $\psi_i(K_2, K_v)$, namely when $i \in \{2, t, 2t\}$, is driven by the fascinating observation that in rare cases such as these the color vector attaining this value is unique and contains only 2 different integers (see Propositions 1, 2, and 3). These values are found using the graph theory technique of amalgamations and may be of particular interest to graph theorists. The results concerning equitable (s, p)-C_4-decompositions of $K_{2v} - F$ then follow as corollaries (see Sect. 4).

It is also worth noting that (s, p)-edge-colorings have the flavor of list edge-colorings. Typically in this area, allowable color lists are assigned to edges, and proper edge-colorings are sought so that each edge receives an allowable color. One could consider a variation where vertices receive allowable color lists, and incident edges must receive a color allowed by the lists of both incident vertices. This can be reformulated as a list edge-coloring problem by assigning the intersection of the lists associated with the two vertices incident with edge e to the allowable list for edge e for each edge e in the graph. This variation does not work well in the sense that finding the smallest p such that any list assignment of size p to each vertex results in a proper s-edge-coloring already requires p to be quite large: p must be greater than $s/2$ just to assure that the resulting intersections assigned to the edges are non-empty. This unpleasing feature can be overcome by adjusting the definition of this proposed variation of list edge-colorings so that, instead of assigning a list of size p to each vertex u from which incident edges must be colored, simply restrict the number of colors on edges incident with u to have size p. These are precisely (s, p)-edge-colorings, and then imposing the equitable requirement simply generalizes the expectation of a proper edge-coloring.

These colorings are also related to those considered by Hilton in [7] and [8]. There, he defined an $(r, r + 1)$-factor to be a spanning subgraph of a graph G in which each vertex has degree r or $r + 1$. Fixing r, he looked for $(r, r + 1)$-factorizations expressing G as the union of edge disjoint $(r, r+1)$-factors. Assigning each $(r, r+1)$-factor a color, an $(r, r+1)$-factorization is exactly an equitable (p, p)-edge-coloring, say E, of G with $b(E, u, i) \in \{r, r + 1\}$ for all $u \in V(G)$ and for all colors i for $1 \leq i \leq p$. He considered a more general problem, considering graphs with degrees within a set of consecutive values. In an effort to do this, he found the possible p for which a regular simple graph, G, can have an equitable (p, p)-edge-coloring, E, with $b(E, u, i) \in \{r, r + 1\}$ for all $u \in V(G)$ and for each color i for $1 \leq i \leq p$. In relation to this paper, Hilton fixed v and $b(E, u, i)$ and found the possible p for which an equitable (p, p)-edge-coloring of K_v exists. Here, we fix p, v (and therefore $b(E, u, i)$), and we focus on the case requiring innovative proof techniques where s must be larger than p in order for an equitable (s, p)-edge-coloring of K_v to exist.

2 Preliminary Results

Lemma 1 ([11]) *Let $v' = 2v = 8tx + 4t + 2$ for some integer x. Let E and E' be an equitable $(2t + 1, 2t)$-edge-coloring and an equitable $(2t + 1, 2t)$- C_4-coloring of K_v and $K_{v'} - F$, respectively. Then*

$$b(E, u, i) = b(E', u', i') = 2x + 1$$

for all $u \in V(K_v)$, $i \in C(E, u)$, $u' \in V(K_{v'} - F)$ and $i' \in C(E', u')$.

In view of Lemma 1, because $b(E, u, i)$ and $b(E'u', i')$ are independent of E, u, i, E', u', and i', we define $b(v) = b'(v') = b(E, u, i) = b(E'u', i') = 2x + 1$. So for such an edge-coloring, at each vertex u, $p = 2t$ colors each occur on exactly $b(v) = 2x + 1$ edges incident with u, and the remaining $s - p = 1$ color appears on no edges incident with u.

Lemma 2, proved in [10], will be used to establish lower bounds for $\psi_i'(K_2, K_v)$ in Lemma 3.

Lemma 2 ([10]) *Let $v \equiv p + 1 \pmod{2p}$. In any equitable (s, p)-edge-coloring E of K_v, for $1 \leq i \leq s$,*

(i) $c_i(E)$ must be even,
(ii) $c_i(E) \geq b(v) + 1 = \frac{v-1}{p} + 1$, and
(iii) if v is odd then $c_i(E) \leq v - 1$.

The following generalizes results in [10] and [11] that considered the $i = 2t + 1$ and $i = 1$ cases respectively. Let $\lceil a \rceil_e$ denote the smallest even integer greater than or equal to a.

Lemma 3 *Let $v = 4tx + 2t + 1$ for some integer x. Then*

$$\psi_i'(K_2, K_v) \geq \max \left\{ b(v) + 1, \left\lceil \frac{v(s-1) - (s-i)(v-1)}{i} \right\rceil_e \right\}$$

for $1 \leq i \leq s$.

Proof Let E be an equitable $(2t + 1, 2t)$-edge-coloring of K_v. Because $p = 2t$ and $s = 2t + 1$, we know each vertex must be missing exactly one color; so $s - 1$ colors appear at each vertex. Counting in two ways, we have $v(s - 1) = \sum_{i=1}^{s} c_i(E)$. By the definition of the color vector, $c_1(E) \leq c_2(E) \leq \cdots \leq c_i(E)$. Also, because v is

Proof Because $s = 2t + 1$ and $p = 2t$, we want every vertex in our final graph, $G = K_v$, to be missing exactly one color. This will be done so that color $2t + 1$ is missing from exactly one vertex and every other color is missing from exactly $2x+1$ vertices. This will show that $\psi'_{2t}(K_2, K_v) \leq 4tx+2t+1-(2x+1) = 4tx+2t-2x$. Then it will follow from Lemma 4 that $\psi'_{2t}(K_2, K_v) = 4tx + 2t - 2x$.

Let $U = \{u_1, \ldots, u_{2t}\}$. We will form the edge-colored graphs G_1, G_2, and G_3 on the vertex set $U \cup \{w\}$ as follows. Once this is done, defining $G' = \bigcup_{i=1}^{3} G_i$, Theorem 2 will be used to detach G' to form $G = K_v$ with our desired coloring. To do so, $\phi : V(G) \to V(G')$ will be the amalgamation function with amalgamation numbers $\eta(w) = 1$ and $\eta(u_i) = 2x + 1$ for $1 \leq i \leq 2t$.

Define α to be a proper $(2t+1)$-edge-coloring of K_{2t+1} on the vertex set $U \cup \{w\}$. Then α is a near-one-factorization. Use colors $\{1, 2, \ldots, 2t + 1\}$ so that each vertex $u_i \in U$ is missing color i and w is missing color $2t + 1$.

(i) Define the $(2t + 1)$-edge-colored graph $G_1 = (2x + 1)K_{2t+1}$ on the vertex set $U \cup \{w\}$ as follows: for each $\{u, u'\} \subseteq U \cup \{w\}$ join u and u' with $2x + 1$ edges colored $\alpha(\{u, u'\})$.

(ii) Define the $(2t + 1)$-edge-colored graph $G_2 = (4x^2 + 2x)K_{2t}$ on the vertex set U as follows: for each $\{u, u'\} \subseteq U$ join u and u' with $(4x^2 + 2x)$ edges colored $\alpha(\{u, u'\})$.

(iii) Define the $(2t + 1)$-edge-colored graph G_3 on the vertex set U by adding $(2x^2 + x)$ loops on u_i of color j if and only if $\alpha(\{w, u_i\}) = j$. (So G_3 has no edges; just loops, each contributing degree 2 to its incident vertex.)

As indicated earlier in the proof, we now define $G' = \bigcup_{i=1}^{3} G_i$. Apply Theorem 2 to G' using the number function η defined earlier to form G. We now show that $G = K_v$ and that the resulting edge-coloring of G has the desired properties.

We first show w is incident in G' to $b(v)$ edges of each color with one exception, a color which appears on no edge, namely color $2t + 1$. We also conclude what this means for G. For $1 \leq j \leq 2t$, by (i), $d_{G'(j)}(w) = 2x + 1 = b(v)$, so $d_{G(j)}(w) = d_{G'(j)}(w)/\eta(w) = b(v)/1 = b(v)$ by Theorem 2 (i). We will now check that each vertex detached from u_i with $1 \leq i \leq 2t$ is incident in G to $b(v)$ edges of each color, with one exception, a color which appears on no edge, namely color i. For $1 \leq i \leq 2t$, $1 \leq j \leq 2t + 1$, and $i \neq j$, by (i) and either (ii) or (iii), $d_{G'(j)}(u_i) = (2x + 1) + (4x^2 + 2x) = 4x^2 + 4x + 1$ or $d_{G'(j)}(u_i) = (2x + 1) + 2(2x^2 + x) = 4x^2 + 4x + 1$ respectively, so $d_{G(j)}(a) = d_{G'(j)}(u_i)/\eta(u_i) = (4x^2 + 4x + 1)/(2x + 1) = 2x + 1 = b(v)$ for each $a \in \phi^{-1}(u_i)$ by Theorem 2 (i).

We will now check that there is exactly one edge between each pair of vertices in G. For $1 \leq i \leq 2t$, by (iii), u_i has $2x^2 + x$ loops, so $m_G(a, a') = \ell_{G'}(u_i)/\binom{\eta(u_i)}{2} = (2x^2 + x)/\left(\frac{(2x+1)(2x)}{2}\right) = 1$ for each distinct $a, a' \in \phi^{-1}(u_i)$ by Theorem 2 (ii). For $1 \leq i, j \leq 2t$ and $i \neq j$, by (i) and (ii), $m_{G'}(u_i, u_j) = (2x + 1) + (4x^2 + 2x) = 4x^2 + 4x + 1$, so $m_G(a, a') = m_{G'}(u_i, u_j)/(\eta(u_i))(\eta(u_j)) = (4x^2 + 4x + 1)/(2x + 1)(2x + 1) = 1$ for each $a \in \phi^{-1}(u_i)$ and $a' \in \phi^{-1}(u_j)$ by Theorem 2 (iii). For $1 \leq i \leq 2t$, by (i), $m_{G'}(w, u_i) = 2x + 1$, so $m_G(a, a') = m_{G'}(w, u_i)/(\eta(w))(\eta(u_i)) =$

$(2x+1)/(1)(2x+1) = 1$ for $a \in \phi^{-1}(w)$ and for each $a' \in \phi^{-1}(u_i)$ by Theorem 2 (iii). Therefore, every vertex in G is attached to each other vertex with exactly one edge. Thus, $G = K_v$. □

The following proposition highlights the interesting property that the color vector used in Theorem 5 to find $\psi_{2t}(K_2, K_v)$ is unique and contains only 2 different integers.

Proposition 1 *Let $v = 4tx + 2t + 1$ for some integer x. In any equitable (s, p)-edge-coloring of K_v with $c_{2t} = \psi'_{2t}(K_2, K_v)$, $c_1 = c_2 = \cdots = c_{2t} = \psi'_{2t}(K_2, K_v)$ and $c_s = v - 1$.*

Proof Since $\psi'_{2t}(K_2, K_v) = 4tx + 2t - 2x$, by definition of color vector and Lemma 2,

$$\sum_{i=1}^{s} c_i \leq 2t(4tx + 2t - 2x) + (v - 1)$$

with equality if and only if

$$c_i = \begin{cases} 4tx + 2t - 2x & \text{for } 1 \leq i \leq 2t, \text{ and} \\ v - 1 & \text{for } i = s. \end{cases}$$

Because $p = 2t$ and $s = 2t + 1$, we know each vertex must be missing exactly one color; so $s - 1$ colors appear at each vertex. Counting in two ways, we have $(s-1)v = \sum_{i=1}^{s} c_i(E)$. The result follows since,

$$\begin{aligned} \sum_{i=1}^{s} c_i &= (s-1)v \\ &= 2t(4tx + 2t + 1) \\ &= 2t(4tx + 2t - 2x) + 4tx + 2t \\ &= 2t(4tx + 2t - 2x) + (v - 1). \end{aligned}$$

□

Lemma 5 *Let $v = 4tx + 2t + 1$ for some integer x. Then, $\psi'_2(K_2, K_v) \geq 2tx + 2t$.*

Proof Using the postulated values of v, s, and i,

$$\begin{aligned} &\left\lceil \frac{v(s-1) - (s-i)(v-1)}{i} \right\rceil_e \\ &\quad = \left\lceil \frac{(4tx + 2t + 1)((2t+1) - 1) - ((2t+1) - 2)((4tx + 2t + 1) - 1)}{2} \right\rceil_e \end{aligned}$$

$$= \left\lceil \frac{4tx + 4t}{2} \right\rceil_e$$

$$= 2tx + 2t.$$

Thus, by Lemma 3, $\psi_2'(K_2, K_v) \geq \max\left\{b(v) + 1, \left\lceil \frac{v(s-1)-(s-i)(v-1)}{i} \right\rceil_e\right\} = 2tx + 2t$. □

Theorem 6 *Let* $v = 4tx + 2t + 1$ *for some integer* x. *Then* $\psi_2'(K_2, K_v) = 2tx + 2t$.

Proof Because $s = 2t + 1$ and $p = 2t$, we want every vertex in our final graph, $G = K_v$, to be missing exactly one color. This will be done so that colors 1 and 2 are each missing from exactly $2tx + 1$ vertices and every other color is missing from exactly one vertex. This will show that $\psi_2'(K_2, K_v) \leq 4tx + 2t + 1 - (2tx + 1) = 2tx + 2t$. Then it will follow from Lemma 5 that $\psi_2'(K_2, K_v) = 2tx + 2t$.

Let $U = \{u_1, u_2\}$, $W = \{w_3, \ldots, w_{2t+1}\}$ and $V = \{v_1, v_2\}$. We will form the edge-colored graphs $G_1, \ldots, G_{10}$ on the vertex set $U \cup W \cup V$ as follows. A general approach is described below similar to the proof used for the next theorem, but also for this theorem a specific example of $G_1, \ldots G_{10}$ follows in Example 1 to potentially help the reader. In (ii) and (iii) below, the vertex $z \notin U \cup W \cup V$ is introduced simply to ensure that w_i is missing color i when G_j $(j \in \{2, 3\})$ is defined. Once this is done, defining $G' = \bigcup_{i=1}^{10} G_i$, Theorem 2 will be used to detach G' to form $G = K_v$ with our desired coloring. To do so, $\phi : V(G) \to V(G')$ will be the amalgamation function with amalgamation numbers $\eta(u_i) = 1$ for $1 \leq i \leq 2$, $\eta(w_i) = 1$ for $3 \leq i \leq 2t + 1$, and $\eta(v_i) = 2tx$ for $1 \leq i \leq 2$.

(i) Define the graph $G_1 = K_{2t+1}$ on the vertex set $U \cup W$. Use a near-one-factorization of G_1 to properly color the edges of G_1 with colors $\{1, 2, \ldots, 2t + 1\}$ so that each vertex $u_i \in U$ is missing color i and each vertex $w_j \in W$ is missing color j.

(ii) Consider $G_2' = (2t - 1)K_{2t-1,1} \cup K_{2t-1,1}$ with parts W and $\{v_1\}$ and W and $\{z\}$ respectively. Using Theorem 4, equitably $(2t, 2t)$-edge-color G_2' with colors $2, \ldots, 2t + 1$ such that the edge $\{w_i, z\}$ is colored i. Form the graph $G_2 = (2tx - x)K_{2t-1,1}$ from G_2' with parts W and $\{v_1\}$ by deleting z and replacing each remaining edge $\{u, v\}$ in G_2' colored j with x edges joining u and v colored j in G_2. (So v_1 is incident with $(2t - 1)x$ edges colored 2 and $(2t - 2)x$ edges of each color $3, 4, \ldots, 2t + 1$.)

(iii) Consider $G_3' = (2t - 1)K_{2t-1,1} \cup K_{2t-1,1}$ with parts W and $\{v_2\}$ and W and $\{z\}$ respectively. Using Theorem 4, equitably $(2t, 2t)$-edge-color G_2' with colors $1, 3, 4, \ldots, 2t + 1$ such that the edge $\{w_i, z\}$ is colored i. Form the graph $G_3 = (2tx - x)K_{2t-1,1}$ from G_3' with parts W and $\{v_2\}$ by deleting z and replacing each remaining edge $\{u, v\}$ in G_3' colored j with x edges joining u and v colored j in G_3. (So v_2 is incident with $(2t - 1)x$ edges colored 1 and $(2t - 2)x$ edges of each color $3, \ldots, 2t + 1$.)

(iv) Define the graph $G_4 = x(K_{2t-1,1})$ with parts W and $\{v_1\}$ with all edges colored 2.

(v) Define the graph $G_5 = x(K_{2t-1,1})$ with parts W and $\{v_2\}$ with all edges colored 1.

(vi) Define the graph G_6 to be the bipartite graph with parts U and V such that u_i and v_j are connected with $2tx - 2x$ edges if $i = j$ and $2tx$ edges otherwise. Using Theorem 4, equitably $(2t-1, 2t-1)$-edge-color G_6 with colors $3, \ldots, 2t+1$. (So, for $1 \leq i \leq 2$ and $3 \leq j \leq 2t+1$, v_i is incident to $((2tx - 2x) + 2tx)/(2t-1) = 2x$ edges of color j.)

(vii) Let $G_7 = (2x)K_{1,1}$ with parts $\{u_1\}$ and $\{v_1\}$ with all edges colored 2.

(viii) Let $G_8 = (2x)K_{1,1}$ with parts $\{u_2\}$ and $\{v_2\}$ with all edges colored 1.

(ix) Let G_9 be formed from $2tx^2 - tx$ loops colored 2 on vertex v_1 and $2tx^2 - tx$ loops colored 1 on vertex v_2.

(x) Let G_{10} be formed from $(4t^2x^2)K_2$ on the vertex set V by adding $2t^2x^2 - 2tx^2$ loops to each vertex. Pair the loops forming edge-disjoint 2-factors (each loop contributes 2 to the degree of its incident vertex) and pair the edges to partition $(4t^2x^2)K_2$ into 2-factors. Therefore, G_{10} is decomposed into $(4t^2x^2 - 2tx^2)$ 2-factors. By appropriately combining the 2-factors, decompose G_{10} into $2t-1$ $(4tx^2)$-factors, coloring their edges with $3, 4, \ldots, 2t+1$ in turn.

As indicated earlier in the proof, we now define $G' = \cup_{i=1}^{10} G_i$. Apply Theorem 2 to G' using the number function η defined earlier to form G. We now show that $G = K_v$ and the resulting edge-coloring of G has the desired properties.

We first show the vertices in the sets U and W are incident in G' to $b(v)$ edges of each color with one exception, a color which appears on no edges, namely color i for each $u_i \in U$ and color j for each $w_j \in W$. We also conclude what this means for G. For $1 \leq i \leq 2$ and $3 \leq j \leq 2t+1$, using (i) and (vi), $d_{G'(j)}(u_i) = 1 + 2x = b(v)$, so $d_{G(j)}(a) = d_{G'(j)}(u_i)/\eta(u_i) = b(v)/1 = b(v)$ for each $a \in \phi^{-1}(u_i)$ by Theorem 2 (i). For $1 \leq i \leq 2$, $1 \leq j \leq 2$, and $i \neq j$, using (i) and (vii) or (viii), $d_{G'(j)}(u_i) = 1 + 2x = b(v)$, so $d_{G(j)}(a) = d_{G'(j)}(u_i)/\eta(u_i) = b(v)/1 = b(v)$ for each $a \in \phi^{-1}(u_i)$ by Theorem 2 (i). For $3 \leq i \leq 2t+1$, $1 \leq j \leq 2t+1$, and $i \neq j$, using (i) and (ii) and (iii) if $3 \leq j \leq 2t+1$, (ii) and (iv) if $j = 2$, or (iii) and (v) if $j = 1$, $d_{G'(j)}(w_i) = 1 + x + x = b(v)$, so $d_{G(j)}(a) = d_{G'(j)}(w_i)/\eta(w_i) = b(v)/1 = b(v)$ for each $a \in \phi^{-1}(w_i)$ by Theorem 2 (i).

We will now check that each vertex detached from v_i with $1 \leq i \leq 2$ is incident in G to $b(v)$ edges of each color with one exception, a color which appears on no edge, namely color i for v_i. For $3 \leq j \leq 2t+1$, using (ii), (vi), and (x), $d_{G'(j)}(v_1) = (2t-2)x + 2x + 4tx^2 = 2tx + 4tx^2$, so $d_{G(j)}(a) = d_{G'(j)}(v_1)/\eta(v_1) = (2tx + 4tx^2)/(2tx) = 1 + 2x = b(v)$ for each $a \in \phi^{-1}(v_1)$ by Theorem 2 (i). Using (ii), (iv), (vii), and (ix), $d_{G'(2)}(v_1) = (2t-1)x + (2t-1)x + 2x + 2(2tx^2 - tx) = 2tx + 4tx^2$, so $d_{G(2)}(a) = d_{G'(2)}(v_1)/\eta(v_1) = (2tx + 4tx^2)/(2tx) = 1 + 2x = b(v)$ for each $a \in \phi^{-1}(v_1)$ by Theorem 2 (i). Similarly, for $3 \leq j \leq 2t+1$, using (iii), (vi), and (x), $d_{G'(j)}(v_2) = (2t-2)x + 2x + 4tx^2 = 2tx + 4tx^2$, so $d_{G(j)}(a) = d_{G'(j)}(v_2)/\eta(v_2) = (2tx + 4tx^2)/(2tx) = 1 + 2x = b(v)$ for each $a \in \phi^{-1}(v_2)$ by Theorem 2 (i). Using (iii), (v), (viii), and (ix), $d_{G'(1)}(v_2) = (2t-1)x + (2t-1)x + 2x + 2(2tx^2 - tx) = 2tx + 4tx^2$, so $d_{G(1)}(a) = d_{G'(1)}(v_2)/\eta(v_2) = (2tx + 4tx^2)/(2tx) = 1 + 2x = b(v)$ for each $a \in \phi^{-1}(v_2)$ by Theorem 2 (i).

$$P + j = (b_1, h_{1,j}, h_{2,j}, \dots h_{(t-1),j}, b_2)$$

where $h_{i,j} = a_{((-1)^i \lfloor \frac{i}{2} \rfloor + j + 1) \pmod t}$. Thus, $\{P + j \mid 0 \le j \le t - 1\} \cup \{C\}$ is the desired decomposition. □

Lemma 8 *Let $v = 4tx + 2t + 1$ for some integer x. Then, $\psi_t'(K_2, K_v) \ge 4tx + 2t - 4x$.*

Proof Using the postulated values of v, s, and i,

$$\begin{aligned} &\left\lceil \frac{v(s-1) - (s-i)(v-1)}{i} \right\rceil_e \\ &= \left\lceil \frac{(4tx + 2t + 1)((2t+1) - 1) - ((2t+1) - t)((4tx + 2t + 1) - 1)}{t} \right\rceil_e \\ &= \left\lceil \frac{4t^2 x + 2t^2 - 4tx}{t} \right\rceil_e \\ &= 4tx + 2t - 4x. \end{aligned}$$

So, using Lemma 3, $\psi_t'(K_2, K_v) \ge \max\left\{b(v) + 1, \left\lceil \frac{v(s-1)-(s-i)(v-1)}{i} \right\rceil_e\right\} \ge 4tx + 2t - 4x$. □

Theorem 8 *Let $v = 4tx + 2t + 1$ for some integer x. Then, $\psi_t'(K_2, K_v) = 4tx + 2t - 4x$.*

Proof Because $s = 2t+1$ and $p = 2t$, we want every vertex in our final graph, $G = K_v$, to be missing exactly one color. This will be done so that colors $t+1, \dots, 2t+1$ are each missing from exactly one vertex and colors $1, \dots, t$ are each missing from exactly $4x+1$ vertices. This will show that $\psi_t'(K_2, K_v) \le 4tx+2t+1-(4x+1) = 4tx+2t-4x$. Then it will follow from Lemma 8 that $\psi_t'(K_2, K_v) = 4tx+2t-4x$.

Let $U = \{u_1, \dots, u_t\}$, $U' = \{u_{t+1}, \dots u_{2t-1}\}$, $W = \{w_{2t}, w_{2t+1}\}$, and $V = \{v_1, \dots, v_t\}$. We will form the edge-colored graphs $G_1, \dots, G_7$ on the vertex set $U \cup U' \cup W \cup V$ as follows. In (ii), (iii), and (iv) below, z_1 and z_2 are introduced simply to ensure that v_j is missing color j and u_i is missing color i respectively when G_k is defined for $k = 2, 3, 4$. Once this is done, defining $G' = \bigcup_{i=1}^7 G_i$, Theorem 2 will be used to detach G' to form $G = K_v$ with our desired coloring. To do so, $\phi : V(G) \to V(G')$ will be the amalgamation function with amalgamation numbers $\eta(u_i) = 1$ for $1 \le i \le 2t - 1$, $\eta(w_i) = 1$ for $2t \le i \le 2t + 1$, and $\eta(v_i) = 4x$ for $1 \le i \le t$.

(i) Define the graph $G_1 = K_{2t+1}$ with vertex set $U \cup U' \cup W$. Use a near-one-factorization of G_1 to properly color the edges of G_1 with colors $\{1, 2, \dots, 2t + 1\}$ so that each vertex $u_i \in U \cup U'$ is missing color i and each vertex $w_j \in W$ is missing color j.

(ii) Let $H = 2K_{t,t}$ with parts U and V, $H' = (t+1)K_2$ on the vertex set $\{z_1, z_2\}$, $H'' = K_{1,t}$ with parts $\{z_1\}$ and V, and $H''' = K_{1,t}$ with parts $\{z_2\}$ and U. Let $G_2' = H \cup H' \cup H'' \cup H'''$. Properly edge-color G_2' with all $2t + 1$ colors such that the $t+1$ edges with endpoints z_1 and z_2 are colored $t+1, \dots 2t+1$. Rename the vertices in $U \cup V$ so that color i is on edge $\{u_i, z_2\}$ and $\{v_i, z_1\}$

for $1 \leq i \leq t$. Form the graph $G_2 = (4x)K_{t,t}$ from G'_2 with parts U and V by deleting z_1 and z_2 and then replacing each remaining edge $\{u, v\}$ in G'_2 colored j with $2x$ edges joining u and v colored j in G_2.

(iii) Let $G'_3 = K_{t+1,t+1}$ with parts $U' \cup W$ and $V \cup \{z_2\}$. Properly edge-color G'_3 with colors $t+1, \ldots, 2t+1$ such that the edge $\{u_i, z_2\}$ is colored i and the edge $\{w_j, z_2\}$ is colored j for each $u_i \in U'$ and $w_j \in W$. Form the graph $G_3 = (2x)K_{t+1,t}$ from G'_3 with parts $U' \cup W$ and V by deleting z_2 and then replacing each remaining edge $\{u, v\}$ in G'_3 colored j with $2x$ edges joining u and v colored j in G_3.

(iv) Let $G'_4 = K_{t,t}$ with parts $U' \cup \{z_1\}$ and V. Properly edge-color G'_4 with colors $1, \ldots, t$ such that the edge $\{z_1, v_i\}$ is colored i. Form the graph $G_4 = (2x)K_{t-1,t}$ from G'_4 with parts U' and V by deleting z_1 and then replacing each remaining edge $\{u, v\}$ in G'_4 colored j with $2x$ edges joining u and v colored j in G_4.

(v) Let $G'_5 = K_2^C \vee 2K_t$ with W being the vertex set of K_2^C and V being the vertex set of the K_t. Use Lemma 7 to decompose G'_5 into a t-cycle on the vertices $v_1, \ldots v_t$ and t paths, each path being of length t with ends w_{2t} and w_{2t+1}. Color the edges of the t-cycle with color $t+1$. Each path P_i, for $1 \leq i \leq t$, is missing exactly one vertex, namely v_i; color the edges of P_i with color i. Define the graph $G_5 = (2x)(K_2^C \vee 2K_t)$ with W being the vertex set of the K_2^C and V being the vertex set of the K_t by replacing each edge $\{u, v\}$ in G'_5 colored j with $2x$ edges joining u and v colored j in G_5.

(vi) Let $G'_6 = 2K_t$ with vertex set V. By Lemma 6, G'_6 can be decomposed into $(t-1)$-cycles, $C_1, \ldots, C_t$, where, for $1 \leq i \leq t$, C_i is missing exactly one vertex, namely v_i. Color the edges of C_i with color i for $1 \leq i \leq t$. Define the graph $G_6 = (8x^2 - 4x)K_t$ with vertex set V by replacing each edge $\{u, v\}$ in G'_6 colored j with $4x^2 - 2x$ edges joining u and v colored j in G_6.

(vii) Let G_7 be formed from $(8x^2)K_t$ on vertex set V by adding $8x^2 - 2x$ loops to each vertex. Partition the loops into edge-disjoint 2-factors (each loop contributes 2 to the degree of its incident vertex) and use Theorem 3 to partition $(8x^2)K_t$ into edge-disjoint 2-factors. Therefore, G_7 is decomposed into $8x^2 - 2x + \frac{8x^2(t-1)}{2} = 4x^2 - 2x + 4x^2 t$ 2-factors. By appropriately combining the 2-factors, decompose G_7 into one $(8x^2 - 4x)$-factor, coloring its edges $(t+1)$, and t $(8x^2)$-factors, coloring their edges $t+2, \ldots, 2t+1$ in turn.

As indicated earlier in the proof, we now define $G' = \bigcup_{i=1}^{7} G_i$. Apply Theorem 2 to G' using the number function η defined earlier to form G. We now show that $G = K_v$ and that the resulting edge-coloring of G has the desired properties.

We first show the vertices in the set U, U', and W are incident in G' to $b(v)$ edges of each color with one exception, a color which appears on no edge, namely color i for each $u_i \in U \cup U'$ and color j for each $w_j \in W$. We also conclude what this means for G. For $1 \leq i \leq t$, $1 \leq j \leq 2t+1$, and $i \neq j$, using (i) and (ii),

$d_{G'(j)}(u_i) = 1 + 2x = b(v)$, so $d_{G(j)}(a) = d_{G'(j)}(u_i)/\eta(u_i) = b(v)/1 = b(v)$ for each $a \in \phi^{-1}(u_i)$ by Theorem 2 (i). For $t+1 \le i \le 2t-1$ and $t+1 \le j \le 2t+1$, using (i) and (iii), $d_{G'(j)}(u_i) = 1 + 2x = b(v)$, so $d_{G(j)}(a) = d_{G'(j)}(u_i)/\eta(u_i) = b(v)/1 = b(v)$ for each $a \in \phi^{-1}(u_i)$ by Theorem 2 (i). For $t+1 \le i \le 2t-1$ and $1 \le j \le t$, using (i) and (iv), $d_{G'(j)}(u_i) = 1 + 2x = b(v)$, so $d_{G(j)}(a) = d_{G'(j)}(u_i)/\eta(u_i) = b(v)/1 = b(v)$ for each $a \in \phi^{-1}(u_i)$ by Theorem 2 (i). For $2t \le i \le 2t+1, t+1 \le j \le 2t+1$, and $i \ne j$, using (i) and (iii), $d_{G'(j)}(w_i) = 1 + 2x = b(v)$, so $d_{G(j)}(a) = d_{G'(j)}(w_i)/\eta(w_i) = b(v)/1 = b(v)$ for each $a \in \phi^{-1}(w_i)$ by Theorem 2 (i). For $2t \le i \le 2t+1$ and $1 \le j \le t$, using (i) and (v), $d_{G'(j)}(w_i) = 1+2x = b(v)$, so $d_{G(j)}(a) = d_{G'(j)}(w_i)/\eta(w_i) = b(v)/1 = b(v)$ for each $a \in \phi^{-1}(w_i)$ by Theorem 2 (i).

We will now check that each vertex detached from v_i with $1 \le i \le t$ is incident in G to $b(v)$ edges of all colors with one exception, a color which appears on no edge, namely color i for each vertex detached from v_i. For $1 \le i \le t$, $1 \le j \le t$, and $i \ne j$, by (ii), (iv), (v), and (vi), $d_{G'(j)}(v_i) = 2x + 2x + 2(2x) + 2(4x^2 - 2x) = 8x^2 + 4x$, so $d_{G(j)}(a) = d_{G'(j)}(v_i)/\eta(v_i) = (8x^2 + 4x)/(4x) = 2x + 1 = b(v)$ for each $a \in \phi^{-1}(v_i)$ by Theorem 2 (i). For $1 \le i \le t$ and $j = t + 1$, by (ii), (iii), (v), and (vii), $d_{G'(j)}(v_i) = 2x + 2x + 2(2x) + (8x^2 - 4x) = 8x^2 + 4x$, so $d_{G(j)}(a) = d_{G'(j)}(v_i)/\eta(v_i) = (8x^2 + 4x)/(4x) = 2x + 1 = b(v)$ for each $a \in \phi^{-1}(v_i)$ by Theorem 2 (i). For $1 \le i \le t$ and $t+2 \le j \le 2t+1$, by (ii), (iii), and (vii), $d_{G'(j)}(v_i) = 2x+2x+8x^2 = 8x^2+4x$, so $d_{G(j)}(a) = d_{G'(j)}(v_i)/\eta(v_i) = (8x^2 + 4x)/(4x) = 2x + 1 = b(v)$ for each $a \in \phi^{-1}(v_i)$ by Theorem 2 (i).

We will now check there is exactly one edge between each pair of vertices in G. By (i), $m_{G'}(b, b') = 1$ for $b, b' \in (U \cup U' \cup W)$, so $m_G(a, a') = m_{G'}(b, b')/(\eta(b)\eta(b')) = 1/((1)(1)) = 1$ for each $a \in \phi^{-1}(b)$ and $a' \in \phi^{-1}(b')$ by Theorem 2 (iii). For each $u \in U$ and $v \in V$, by (ii), $m_{G'}(u, v) = 4x$, so $m_G(a, a') = m_{G'}(u, v)/(\eta(u)\eta(v)) = 4x/((1)(4x)) = 1$ for each $a \in \phi^{-1}(u)$ and $a' \in \phi^{-1}(v)$ by Theorem 2 (iii). For $u' \in U'$ and $v \in V$, by (iii) and (iv), $m_{G'}(u', v) = 2x + 2x = 4x$, so $m_G(a, a') = m_{G'}(u', v)/(\eta(u')\eta(v)) = 4x/((1)(4x)) = 1$ for each $a \in \phi^{-1}(u')$ and $a' \in \phi^{-1}(v)$ by Theorem 2 (iii). For $w \in W$ and $v \in V$, by (iii) and (v), $m_{G'}(w, v) = 2x+2x = 4x$, so $m_G(a, a') = m_{G'}(w, v)/(\eta(w)\eta(v)) = 4x/((1)(4x)) = 1$ for each $a \in \phi^{-1}(w)$ and $a' \in \phi^{-1}(v)$ by Theorem 2 (iii). For $v, v' \in V$, by (v), (vi), and (vii), $m_{G'}(v, v') = 4x + (8x^2 - 4x) + 8x^2 = 16x^2$, so $m_G(a, a') = m_{G'}(v, v')/(\eta(v)\eta(v')) = 16x^2/((4x)(4x)) = 1$ for each $a \in \phi^{-1}(v)$ and $a' \in \phi^{-1}(v')$ by Theorem 2 (iii). Finally, for $1 \le i \le t$ by (vii), v_i has $8x^2 - 2x$ loops, so $m_G(a, a') = \ell_{G'}(v_i)/\binom{\eta(v_i)}{2} = (8x^2 - 2x)/\left(\frac{(4x)(4x-1)}{2}\right) = 1$ for each distinct $a, a' \in \phi^{-1}(v_i)$ by Theorem 2 (ii). Therefore, every vertex in G is attached to each other vertex with exactly one edge. Thus, $G = K_v$. □

Proposition 3 *Let $v = 4tx + 2t + 1$ for some integer x. In any equitable (s, p)-edge-coloring of K_v with $c_t = \psi_t'(K_2, K_v)$, $c_1 = c_2 = \cdots = c_t = \psi_t'(K_2, K_v)$ and $c_{t+1} = c_{t+2} = \cdots = c_s = v - 1$.*

Proof Since $\psi_t'(K_2, K_v) = 4tx + 2t - 4x$, by definition of color vector and Lemma 2,

$$\sum_{i=1}^{s} c_i \leq t(4tx + 2t - 4x) + (s - t)(v - 1)$$

with equality if and only if

$$c_i = \begin{cases} 4tx + 2t - 4x & \text{for } 1 \leq i \leq t, \text{ and} \\ v - 1 & \text{for } t + 1 \leq i \leq s. \end{cases}$$

Because $p = 2t$ and $s = 2t + 1$, we know each vertex must be missing exactly one color; so $s - 1$ colors appear at each vertex. Counting in two ways, we have $v(s - 1) = \sum_{i=1}^{s} c_i(E)$. The result follows since,

$$\begin{aligned} \sum_{i=1}^{s} c_i &= (s - 1)v \\ &= 2t(4tx + 2t + 1) \\ &= t(4tx + 2t - 4x) + t(4tx + 2t) + 4xt + 2t \\ &= t(4tx + 2t - 4x) + (t + 1)(4tx + 2t) \\ &= t(4tx + 2t - 4x) + (s - t)(v - 1). \end{aligned}$$

□

The following corollary of our main theorems obtains additional values $\psi'_i(K_2, K_v)$.

Corollary 1 *If $i < j$ and*

$$\max\left\{b(v) + 1, \left\lceil \frac{v(s - 1) - (s - i)(v - 1)}{i} \right\rceil_e \right\} = \psi'_j(K_2, K_v),$$

then $\psi'_i(K_2, K_v) = \psi'_j(K_2, K_v)$.

Proof By Lemma 3, $\psi'_i(K_2, K_v) \geq \max\left\{b(v) + 1, \left\lceil \frac{v(s-1)-(s-i)(v-1)}{i} \right\rceil_e \right\}$. Using assumption, $\psi'_i(K_2, K_v) \geq \psi'_j(K_2, K_v)$. By definition of color vector, $\psi'_i(K_2, K_v) \leq \psi'_j(K_2, K_v)$. Thus, $\psi'_i(K_2, K_v) = \psi'_j(K_2, K_v)$. □

The significance of the previous corollary is demonstrated in the following example. It is extremely useful when $j > t$ which is seen in Example 2 (ii).

Example 2

(i) Let $x = 1$ and $t = 5$. Thus, $v = 4tx + 2t + 1 = 31$. By Theorem 8, $\psi'_5(K_2, K_v) = 4tx + 2t - 4x = 26$. Also, for $i = 4$,

$$\begin{aligned}
&\left\lceil \frac{v'(s-1)-(s-i)(v'-2)}{i} \right\rceil_{d:4} \\
&\quad = \left\lceil \frac{(8tx+4t+2)((2t+1)-1)-((2t+1)-2)((8tx+4t+2)-2)}{2} \right\rceil_{d:4} \\
&\quad = \left\lceil \frac{8tx+8t}{2} \right\rceil_{d:4} \\
&\quad = 4tx+4t.
\end{aligned}$$

By Lemma 11, $\psi_2'(C_4, K_{v'}-F) \geq 4tx+4t$. Thus, $\psi_2'(C_4, K_{v'}-F) = 4tx+4t$. □

Next, we will find $\psi_t'(C_4, K_{v'}-F)$. We first note that the case where $t=1$ has been settled since $\psi_1'(C_4, K_{v'}-F)$ was found in [11], as stated below.

Theorem 11 ([11]) *Let $v' = 8tx+4t+2$ for some integer x. Then, $\psi_1'(C_4, K_{v'}-F) = \max\{2(b'(v')+1), 4t\}$.*

Theorem 12 *Let $v' = 8tx+4t+2$ for some integer x. Assume $t \geq 2$. Then, $\psi_t'(C_4, K_{v'}-F) = 8tx+4t-8x$.*

Proof By Lemma 9 and Theorem 8, $\psi_t'(C_4, K_{v'}-F) \leq 2(4tx+2t-4x)$. Also, using the postulated values of v', s, and i,

$$\begin{aligned}
&\left\lceil \frac{v'(s-1)-(s-i)(v'-2)}{i} \right\rceil_{d:4} \\
&\quad = \left\lceil \frac{(8tx+4t+2)((2t+1)-1)-((2t+1)-t)((8tx+4t+2)-2)}{t} \right\rceil_{d:4} \\
&\quad = \left\lceil \frac{8t^2x+4t^2-8tx}{t} \right\rceil_{d:4} \\
&\quad = 8tx+4t-8x.
\end{aligned}$$

By Lemma 11, $\psi_{2t}'(C_4, K_{v'}-F) \geq 8tx+4t-8x$. Thus, $\psi_t'(C_4, K_{v'}-F) = 8tx+4t-8x$. □

Proposition 4 *Let $v' = 8tx+4t+2$ for some integer x. In any equitable (s,p)-C_4-coloring of $K_{v'}$ with $c_i = \psi_i'(K_2, K_v)$ for $i \in \{2, t, 2t\}$, $c_1 = c_2 = \cdots = c_i = \psi_i'(K_2, K_v)$ and $c_{i+1} = c_{i+2} = \cdots = c_s = v-1$.*

The proof of the previous proposition is similar to the proofs of Propositions 1, 2, and 3.

Corollary 2 *If $i < j$ and $\max\left\{2(b'(v')+1), \left\lceil \frac{v'(s-1)-(s-i)(v'-2)}{i} \right\rceil_{d:4}\right\} = \psi_j'(C_4, K_{v'})$, then $\psi_i'(C_4, K_{v'}) = \psi_j'(C_4, K_{v'})$.*

The proof of the previous corollary is similar to the proof of Corollary 1.

References

1. M.A. Bahmanian and C.A. Rodger. Multiply balanced edge colorings of multigraphs. J. Graph Theory **70**(3), 297–317 (2012)
2. C.J. Colbourn, A. Rosa, Specialized block-colourings of Steiner triple systems and the upper chromatic index. Graphs Combin. **19**(3), 335–345 (2003)
3. D. deWerra, Equitable colorations of graphs. Rev. Française Informat. Recherche Opérationnelle **5**(Sér. R-3), 3–8 (1971)
4. M. Gionfriddo, G. Ragusa, Equitable specialized block-colourings for 4-cycle systems—II. Discrete Math. **310**(13–14), 1986–1994 (2010)
5. M. Gionfriddo, P. Horák, L. Milazzo, A. Rosa, Equitable specialized block-colourings for Steiner triple systems. Graphs Combin. **24**(4), 313–326 (2008)
6. L. Gionfriddo, M. Gionfriddo, G. Ragusa, Equitable specialized block-colourings for 4-cycle systems—I. Discrete Math. **310**(22), 3126–3131 (2010)
7. A.J.W. Hilton, $(r, r+1)$-factorizations of $(d, d+1)$-graphs. Discrete Math. **308**(5–6), 645–669 (2008)
8. A.J.W. Hilton, On the number of $(r, r+1)$-factors in an $(r, r+1)$-factorization of a simple graph. J. Graph Theory **60**(4), 257–268 (2009)
9. S. Li, C.A. Rodger, Equitable block-colorings of C_4-decompositions of $K_v - F$. Discrete Math. **339**(5), 1519–1524 (2016)
10. S. Li, E.B. Matson, C.A. Rodger, Extreme equitable block-colorings of C_4-decompositions of $K_v - F$. Australas. J. Combin. **71**, 92–103 (2018)
11. E.B. Matson, C.A. Rodger, More extreme equitable colorings of decompositions of K_v and $K_v - F$. Discrete Math. **341**(4), 1178–1184 (2018)

On Maximum Packings of the λ-Fold Complete 3-Uniform Hypergraph with Loose 4-Paths

Ryan C. Bunge, Lisa Hamanaka, Cody Hatzer, Tracer Mills, and Lauren Nowak

Abstract A 3-uniform loose 4-path is a hypergraph with vertex set $\{v_1, v_2, v_3, v_4, v_5, v_6, v_7, v_8, v_9\}$ and with edge set $\big\{\{v_1, v_2, v_3\}, \{v_3, v_4, v_5\}, \{v_5, v_6, v_7\}, \{v_7, v_8, v_9\}\big\}$. For all positive integers λ and v, we find a maximum packing with 3-uniform loose 4-paths of the λ-fold complete 3-uniform hypergraph of order v. We show that, if $v \geq 9$, such a packing has a leave of 3 or fewer edges.

Keywords Spectrum problem · Maximum packing · λ-fold complete 3-uniform hypergraph

MSC Codes 05C65

1 Introduction

A *hypergraph* consists of a finite, nonempty set of *vertices* and a finite collection of nonempty subsets of the vertices called *hyperedges* or simply *edges*. For a given hypergraph H, we use $V(H)$ and $E(H)$ to denote the vertex set and the edge set (or multiset) of H, respectively. We call $|V(H)|$ and $|E(H)|$ the *order* and *size* of H, respectively. A hypergraph H is *simple* if no edge appears more than once in $E(H)$.

In Honor of Professor Saad El-Zanati on the Occasion of his Retirement from Illinois State University

---

R. C. Bunge (✉) · C. Hatzer · T. Mills
Illinois State University, Normal, IL, USA
e-mail: rcbunge@ilstu.edu

L. Hamanaka
Iowa State University, Ames, IA, USA

L. Nowak
Leadership High School, San Francisco, CA, USA

A. Bahmanian (ed.), *Combinatorics and Combinatorial Computing*, Springer Proceedings in Mathematics & Statistics 516,
https://doi.org/10.1007/978-3-032-08549-8_4

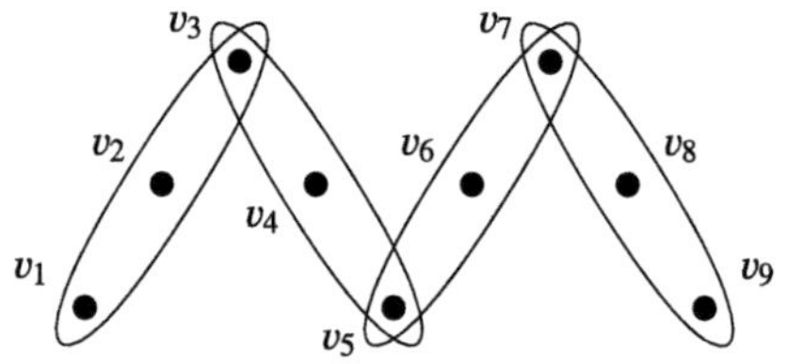

Fig. 1 The loose 3-uniform 4-path, or $LP_4^{(3)}$, denoted by $H[v_1, v_2, v_3, v_4, v_5, v_6, v_7, v_8, v_9]$

1.1 Additional Notation and Terminology

Let $\mathbb{Z}_n$ denote the group of integers modulo n. We next define some notation for certain types of 3-uniform hypergraphs.

Let U_1, U_2, U_3 be pairwise disjoint sets. The hypergraph with vertex set $U_1 \cup U_2 \cup U_3$ and edge set consisting of all 3-element sets having exactly one vertex in each of U_1, U_2, U_3 is denoted by $K^{(3)}_{U_1,U_2,U_3}$. The hypergraph with vertex set $U_1 \cup U_2$ and edge set consisting of all 3-element sets having at most 2 vertices in each of U_1, U_2 is denoted by $L^{(3)}_{U_1,U_2}$. If $|U_i| = u_i$ for $i \in \{1, 2, 3\}$, we may use $K^{(3)}_{u_1,u_2,u_3}$ or $L^{(3)}_{u_1,u_2}$ to denote any hypergraph that is isomorphic to $K^{(3)}_{U_1,U_2,U_3}$ or $L^{(3)}_{U_1,U_2}$, respectively.

2 Main Results

We begin by giving necessary conditions for the existence of an $LP_4^{(4)}$-decomposition of ${}^{\lambda}K_v^{(3)}$. Some obvious necessary conditions include that the size of ${}^{\lambda}K_v^{(3)}$ must be divisible by the size of $LP_4^{(3)}$ and that v must be at least the order of $LP_4^{(3)}$. Thus we have the following.

Lemma 1 *Let $v \geq 9$ be an integer. If there exists an $LP_4^{(3)}$-decomposition of λ-fold $K_v^{(3)}$, then one of the following must hold:*

- $\gcd(\lambda, 4) = 1$ *and* $v \equiv 0, 1, 2, 4,$ *or* $6 \pmod 8$*;*
- $\gcd(\lambda, 4) = 2$ *and* $v \equiv 0, 1,$ *or* $2 \pmod 4$*;*
- $\gcd(\lambda, 4) = 4$.

Proof Suppose there exists an $LP_4^{(3)}$-decomposition of λ-fold $K_v^{(3)}$. Since we have $\left|E\left(LP_4^{(3)}\right)\right| = 4$, it follows that $4 \mid \lambda\binom{v}{3} = \lambda v(v-1)(v-2)/6$, and thus we have that $8 \mid \lambda v(v-1)(v-2)$. First, if $\gcd(\lambda, 4) = 1$, then $8 \mid v(v-1)(v-2)$, and thus $v \equiv 0, 1, 2, 4$, or $6 \pmod 8$. Second, if $\gcd(\lambda, 4) = 2$, then $4 \mid v(v-1)(v-2)$, and thus $v \equiv 0, 1$, or $2 \pmod 4$. Finally, if $\gcd(\lambda, 4) = 4$, then $2 \mid v(v-1)(v-2)$, which is true for any $v \geq 9$. □

In what follows, we show how to construct $LP_4^{(3)}$-decompositions of $K_v^{(3)}$ for all $v \equiv 0, 1, 2, 4$, or $6 \pmod 8$ with $v \geq 9$ and maximum $LP_4^{(3)}$-packings of $K_v^{(3)}$ for the remaining values of v. We then demonstrate how to get maximum $LP_4^{(3)}$-packings when $\lambda \geq 2$. Our constructions are dependent on the many small examples given in the Appendix. First, we prove a lemma that is fundamental to our constructions.

Lemma 2 *Let n, x, and r be nonnegative integers such that $nx+r \geq 3$. There exists a decomposition of $K_{nx+r}^{(3)}$ that is comprised of isomorphic copies of the following under the given conditions:*

- $K_r^{(3)}$ *if* $x = 0$,
- $K_{n+r}^{(3)}$ *if* $x \geq 1$,
- $K_{n+r}^{(3)} \setminus K_r^{(3)}$ *if* $x \geq 2$,
- $K_{r,n,n}^{(3)} \cup L_{n,n}^{(3)}$ *if* $x \geq 2$,
- $K_{n,n,n}^{(3)}$ *if* $x \geq 3$.

Furthermore, if $x \geq 1$ and $r \geq 3$, then the decomposition contains exactly one copy of $K_{n+r}^{(3)}$.

Proof If $x \in \{0, 1\}$, the decomposition is trivial. Similarly, if $n = 0$, then $r \geq 3$, and the result is trivial because $K_r^{(3)} = K_{n+r}^{(3)} = K_{nx+r}^{(3)}$ while $K_{n+r}^{(3)} \setminus K_r^{(3)}$, $K_{r,n,n}^{(3)} \cup L_{n,n}^{(3)}$, and $K_{n,n,n}^{(3)}$ are all empty (i.e., contain no edges). For the remainder of the proof, we assume that $x \geq 2$ and $n \geq 1$.

Let $V_0, V_1, \ldots, V_x$ be pairwise disjoint sets of vertices with $|V_0| = r$ and $|V_1| = |V_2| = \cdots = |V_x| = n$. Then, the decomposition of $K_{nx+r}^{(3)}$ results from the fact that the complete 3-uniform hypergraph on the vertex set $V_0 \cup V_1 \cup \cdots \cup V_x$, which is $nx + r$ vertices, can be viewed as the (edge-disjoint) union

$$K_{V_1 \cup V_0}^{(3)} \cup \bigcup_{2 \leq i \leq x} \left(K_{V_i \cup V_0}^{(3)} \setminus K_{V_0}^{(3)} \right) \cup \bigcup_{1 \leq i < j \leq x} \left(K_{V_0, V_i, V_j}^{(3)} \cup L_{V_i, V_j}^{(3)} \right) \cup \bigcup_{1 \leq i < j < k \leq x} \left(K_{V_i, V_j, V_k}^{(3)} \right).$$

In addition, if $r \geq 3$, the single isomorphic copy of $K_{n+r}^{(3)}$ in the decomposition is $K_{V_1 \cup V_0}^{(3)}$. □

An example of how the vertex set in the proof of Lemma 2 is partitioned can be seen in Fig. 2.

We now give our main results on $LP_4^{(3)}$-decompositions of onefold $K_v^{(3)}$.

Theorem 1 *There exists an $LP_4^{(3)}$-decomposition of $K_v^{(3)}$ if and only if $v \equiv 0, 1, 2, 4$, or $6 \pmod 8$ and $v \notin \{4, 6, 8\}$.*

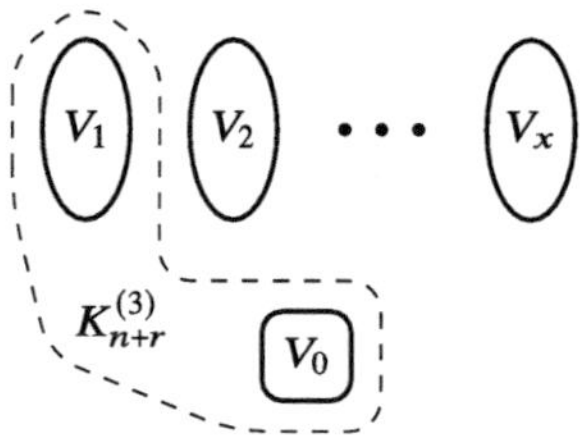

Fig. 2 Vertex partition of $K^{(3)}_{nx+r}$ as described in the proof for Lemma 2. Recall that V_0 contains r vertices, while all other V_i each contain n vetices. The sole copy of $K^{(3)}_{n+r}$ is implied with the dashed outlined. All other such groupings of V_0 with a single V_i (with $i \neq 1$) would imply a copy of $K^{(3)}_{n+r} \setminus K^{(3)}_r$. The copies of the complete tripartite hypergraphs $K^{(3)}_{r,n,n}$ and $K^{(3)}_{n,n,n}$ and the copies of the hypergraph $L^{(3)}n$, n can be similarly implied with other groupings of vertices

Proof The necessary conditions for the existence of an $LP^{(3)}_4$-decomposition of $K^{(3)}_v$ are established in Lemma 1. Thus we need only to establish their sufficiency. Let $v = 8x + r$ where $x \geq 1$ and $r \in \{1, 2, 4, 6, 8\}$. By Lemma 2 it suffices to find $LP^{(3)}_4$-decompositions of $K^{(3)}_{8+r}$, $K^{(3)}_{8+r} \setminus K^{(3)}_r$, $K^{(3)}_{r,8,8} \cup L^{(3)}_{8,8}$, and $K^{(3)}_{8,8,8}$. We note that if $r \in \{1, 2\}$ then $K^{(3)}_{8+r} \setminus K^{(3)}_r$ is isomorphic to $K^{(3)}_{8+r}$. Also, $K^{(3)}_{2,8,8}$ decomposes $K^{(3)}_{4,8,8}$, $K^{(3)}_{6,8,8}$, and $K^{(3)}_{8,8,8}$. Thus, it suffices to find $LP^{(3)}_4$-decompositions of $K^{(3)}_9$, $K^{(3)}_{10}$, $K^{(3)}_{12}$, $K^{(3)}_{14}$, $K^{(3)}_{16}$, $K^{(3)}_{12} \setminus K^{(3)}_4$, $K^{(3)}_{14} \setminus K^{(3)}_6$, $K^{(3)}_{16} \setminus K^{(3)}_8$, $K^{(3)}_{1,8,8} \cup L^{(3)}_{8,8}$, $K^{(3)}_{2,8,8}$, and $L^{(3)}_{8,8}$, which are each shown to exist in Examples 1–8, 10, 12, and 14. □

For values of $v \geq 9$ not in the spectrum mentioned in Theorem 1, we show maximum $LP^{(3)}_4$-packings of onefold $K^{(3)}_v$ with a loose-path leave.

Theorem 2 *If $v \equiv 3, 5,$ or $7 \pmod 8$ and $v \geq 11$, then there exists a maximum $LP^{(3)}_4$-packing of $K^{(3)}_v$ with a leave isomorphic to $LP^{(3)}_n$ where $n \in \{1, 2, 3\}$.*

Proof Let $v = 8x + r$ where $x \geq 1$ and $r \in \{3, 5, 7\}$. By Lemma 2 it suffices to find

- a maximum $LP^{(3)}_4$-packing of $K^{(3)}_{8+r}$ with an $LP^{(3)}_n$ leave and
- $LP^{(3)}_4$-decompositions of $K^{(3)}_{8+r} \setminus K^{(3)}_r$, $K^{(3)}_{r,8,8} \cup L^{(3)}_{8,8}$, and $K^{(3)}_{8,8,8}$.

We note that an $LP^{(3)}_4$-decomposition of $K^{(3)}_{11} \setminus K^{(3)}_3$ is equivalently a maximum $LP^{(3)}_4$-packing of $K^{(3)}_{11}$ with a leave consisting of a single edge, i.e., an $LP^{(3)}_1$. Also, both $K^{(3)}_{5,8,8}$ and $K^{(3)}_{7,8,8}$ can be decomposed into isomorphic copies of $K^{(3)}_{2,8,8}$ and $K^{(3)}_{3,8,8}$. Thus, it suffices to find maximum $LP^{(3)}_4$-packings of $K^{(3)}_{11}$, $K^{(3)}_{13}$, and $K^{(3)}_{15}$ each with a leave consisting of edges in an $LP^{(3)}_n$ with $n \in \{1, 2, 3\}$, which are shown to exist in Examples 15–17, along with $LP^{(3)}_4$-decompositions of $K^{(3)}_{13} \setminus K^{(3)}_5$, $K^{(3)}_{15} \setminus K^{(3)}_7$, $K^{(3)}_{2,8,8}$, $K^{(3)}_{3,8,8}$, and $L^{(3)}_{8,8}$, which are shown to exist in Examples 11, 13, 8, 9, and 6, respectively. □

Lemma 3 *Let λ_1, λ_2, m, n_1, and n_2 be positive integers with $m > \max\{n_1, n_2\}$ and let K be a vertex-transitive, 3-uniform hypergraph with at least $2m + 1$ vertices. If λ_i-fold K admits a maximum $LP_m^{(3)}$-packing where the leave is a loose n_i-path for $i \in \{1, 2\}$, then $(\lambda_1 + \lambda_2)$-fold K admits a maximum $LP_m^{(3)}$-packing where the leave is a loose path with fewer than m edges.*

Proof Let $V(K) = \mathbb{Z}_M$ and, for $i \in \{1, 2\}$, let Δ_i be a maximum $LP_4^{(3)}$-packing of λ_i-fold K with leave L_i. Since K is vertex-transitive, we may assume (without loss of generality) that L_1 and L_2 have as few vertices in common as needed. Specifically, let

$$E(L_1) = \big\{\{0, 1, 2\}, \{2, 3, 4\}, \{4, 5, 6\}, \ldots, \{2n_1 - 2,\ 2n_1 - 1,\ 2n_1\}\big\},$$
$$E(L_2) = \big\{\{2n_1,\ 2n_1 + 1,\ 2n_1 + 2\}, \{2n_1 + 2,\ 2n_1 + 3,\ 2n_1 + 4\},$$
$$\{2n_1 + 4,\ 2n_1 + 5,\ 2n_1 + 6\}, \ldots,$$
$$\{2n_1 + 2n_2 - 2,\ 2n_1 + 2n_2 - 1,\ 2n_1 + 2n_2\}\big\}$$

and let L be the (multi-)hypergraph with edge (multi-)set $E(L_1) \cup E(L_2)$.

In the case where $n_1 + n_2 < m$, we note that L is a loose $(n_1 + n_2)$-path, and

$$\big(\Delta_1 \setminus \{L_1\}\big) \cup \big(\Delta_2 \setminus \{L_2\}\big) \cup \{L\}$$

is a maximum $LP_m^{(3)}$-packing of $(\lambda_1 + \lambda_2)$-fold K where the leave is a loose path with fewer than m edges. In the case where $n_1 + n_2 \geq m$, we note that L has a loose m-path as a subgraph; specifically, the hypergraph with edge set

$$\big\{\{0, 1, 2\}, \{2, 3, 4\}, \{4, 5, 6\}, \ldots, \{2m - 2,\ 2m - 1,\ 2m\}\big\}$$

is a subgraph of L and is a loose m-path because $2m + 1 \leq M$. Let L' be the loose m-path subgraph of L described above and let L'' denote the hypergraph $L \setminus L'$. We note that L'' is a loose path with $n_1 + n_2 - m$ edges, which is fewer than m edges since $m > \max\{n_1, n_2\}$. Finally,

$$\big(\Delta_1 \setminus \{L_1\}\big) \cup \big(\Delta_2 \setminus \{L_2\}\big) \cup \{L', L''\}$$

is a maximum $LP_m^{(3)}$-packing of $(\lambda_1 + \lambda_2)$-fold K where the leave is a loose path with fewer than m edges. □

As a brief example, consider finding a maximum packing of twofold $K_{15}^{(3)}$. By Example 17, we know that (onefold) $K_{15}^{(3)}$ admits a maximum $LP_4^{(3)}$-packing with a loose 3-path as its leave. Since $K_{15}^{(3)}$ is vertex transitive, we may assume two (unique) such packings: one with the leave $\big\{\{0, 1, 2\}, \{2, 3, 4\}, \{4, 5, 6\}\big\}$ and one with the leave $\big\{\{6, 7, 8\}, \{8, 9, 10\}, \{10, 11, 12\}\big\}$ (where we assume $\mathbb{Z}_{15}$ is the

$$B_2 = \{H[0, 8, 1, 9, 2, 10, 3, 11, 4],\ H[0, 6, 1, 7, 2, 8, 3, 9, 4],\\ H[0, 5, 1, 6, 2, 7, 3, 8, 4]\}.$$

Then an $LP_4^{(3)}$-decomposition of $K_{16}^{(3)}$ consists of the orbits of the H-blocks in B_1 under the action of the map $j \mapsto j + 1 \pmod{16}$ along with the orbits of the H-blocks in B_2 under the action of the map $j \mapsto j + 4 \pmod{16}$.

Example 6 Let $V\left(L_{8,8}^{(3)}\right) = \mathbb{Z}_{16}$ with vertex partition $\big\{\{0, 2, 4, 6, 8, 10, 12, 14\}, \{1, 3, 5, 7, 9, 11, 13, 15\}\big\}$ and let

$$B = \{H[0, 10, 5, 8, 1, 3, 14, 7, 12],\ H[3, 13, 2, 5, 0, 4, 11, 14, 6],\\ H[7, 14, 9, 4, 0, 10, 3, 5, 12],\ H[4, 5, 3, 0, 6, 13, 9, 8, 2],\\ H[0, 8, 3, 6, 12, 13, 10, 14, 15],\ H[0, 1, 8, 9, 14, 15, 11, 10, 3],\\ H[0, 1, 3, 4, 7, 12, 8, 9, 15]\}.$$

Then an $LP_4^{(3)}$-decomposition of $L_{8,8}^{(3)}$ consists of the orbits of the H-blocks in B under the action of the map $j \mapsto j + 1 \pmod{16}$.

Example 7 Let $V\left(L_{8,8}^{(3)} \cup K_{1,8,8}^{(3)}\right) = \mathbb{Z}_{16} \cup \{\infty\}$ with vertex partition $\big\{\{\infty\}, \{0, 2, 4, 6, 8, 10, 12, 14\}, \{1, 3, 5, 7, 9, 11, 13, 15\}\big\}$ and let

$$B = \{H[0, 10, 5, 8, 1, 3, 14, 7, 12],\ H[3, 13, 2, 5, 0, 4, 11, 14, 6],\\ H[7, 14, 9, 4, 0, 10, 3, 5, 12],\ H[4, 5, 3, 0, 6, 13, 9, 8, 2],\\ H[0, 8, 3, 6, 12, 13, 10, 14, 15],\ H[0, 1, 8, 9, 14, 15, 11, 10, 3],\\ H[0, 1, 3, 4, 7, 8, \infty, 12, 15],\ H[0, 1, 5, 12, 6, 11, \infty, 3, 10]\}.$$

Then an $LP_4^{(3)}$-decomposition of $L_{8,8}^{(3)} \cup K_{1,8,8}^{(3)}$ consists of the orbits of the H-blocks in B under the action of the map $\infty \mapsto \infty$ and $j \mapsto j + 1 \pmod{16}$.

Example 8 Let $V\left(K_{2,8,8}^{(3)}\right) = \mathbb{Z}_{16} \cup \{\infty_1, \infty_2\}$ with vertex partition $\big\{\{\infty_1, \infty_2\}, \{0, 2, 4, 6, 8, 10, 12, 14\}, \{1, 3, 5, 7, 9, 11, 13, 15\}\big\}$ and let

$$B = \{H[0, 1, \infty_1, 2, 5, 10, \infty_2, 4, 11],\ H[0, 1, \infty_2, 2, 5, 10, \infty_1, 4, 11]\}.$$

Then an $LP_4^{(3)}$-decomposition of $K_{2,8,8}^{(3)}$ consists of the orbits of the H-blocks in B under the action of the map $\infty_i \mapsto \infty_i$, for $i \in \{1, 2\}$, and $j \mapsto j + 1 \pmod{16}$.

Example 9 Let $V\left(K_{3,8,8}^{(3)}\right) = \mathbb{Z}_{16} \cup \{\infty_1, \infty_2, \infty_3\}$ with vertex partition $\big\{\{\infty_1, \infty_2, \infty_3\}, \{0, 2, 4, 6, 8, 10, 12, 14\}, \{1, 3, 5, 7, 9, 11, 13, 15\}\big\}$ and let

$$B = \{H[0, 1, \infty_1, 2, 5, 10, \infty_2, 4, 11],\ H[0, 1, \infty_2, 2, 5, 10, \infty_3, 4, 11],$$
$$H[0, 1, \infty_3, 2, 5, 10, \infty_1, 4, 11]\}.$$

Then an $LP_4^{(3)}$-decomposition of $K_{3,8,8}^{(3)}$ consists of the orbits of the H-blocks in B under the action of the map $\infty_i \mapsto \infty_i$, for $i \in \{1, 2, 3\}$, and $j \mapsto j+1 \pmod{16}$.

Example 10 Let $V\left(K_{12}^{(3)} \setminus K_4^{(3)}\right) = \mathbb{Z}_{12}$ with $0, 3, 6, 9$ being the vertices in the hole and let

$$B_1 = \{H[0, 3, 8, 6, 4, 1, 2, 5, 9],\ H[5, 10, 9, 6, 8, 0, 2, 4, 7],$$
$$H[9, 10, 11, 3, 2, 8, 1, 0, 4],\ H[0, 9, 2, 8, 4, 6, 11, 10, 3]\},$$

$$B_2 = \{H[9, 10, 3, 2, 8, 0, 4, 1, 7],\ H[10, 11, 4, 3, 9, 1, 5, 2, 8],$$
$$H[5, 11, 0, 1, 6, 2, 10, 4, 7],\ H[4, 10, 5, 8, 11, 3, 7, 1, 2],$$
$$H[2, 8, 11, 5, 6, 0, 7, 10, 1],\ H[5, 11, 2, 9, 8, 7, 1, 10, 4]\}.$$

Then an $LP_4^{(3)}$-decomposition of $K_{12}^{(3)} \setminus K_4^{(3)}$ consists of the orbits of the H-blocks in B_1 under the action of the map $j \mapsto j+1 \pmod{12}$ along with the H-blocks in B_2.

Example 11 Let $V\left(K_{13}^{(3)} \setminus K_5^{(3)}\right) = \mathbb{Z}_{12} \cup \{\infty\}$ with $0, 3, 6, 9, \infty$ being the vertices in the hole and let

$$B_1 = \{H[2, 8, 0, 7, 3, 6, 11, 10, 5],\ H[5, 6, 8, 4, 7, 2, 0, 10, 3],$$
$$H[2, 4, 0, 1, 10, 9, 5, 6, 11],\ H[3, 11, 0, 4, \infty, 5, 10, 8, 2],$$
$$H[0, 1, \infty, 2, 4, 5, 6, 8, 3]\},$$

$$B_2 = \{H[7, 10, 4, 1, \infty, 5, 2, 8, 11],\ H[9, 2, 10, 1, 7, 4, \infty, 5, 8],$$
$$H[1, 4, 10, 7, \infty, 8, 11, 2, 5],\ H[4, 7, 1, 10, \infty, 11, 2, 5, 8],$$
$$H[5, 0, 1, 6, 2, \infty, 8, 3, 4],\ H[8, 9, 1, \infty, 7, 2, 3, 10, 11],$$
$$H[5, 8, 11, 0, 4, \infty, 10, 2, 6],\ H[0, 8, 4, 9, 5, 10, 6, 7, 11],$$
$$H[1, 9, 5, \infty, 11, 3, 7, 8, 0]\}.$$

Then an $LP_4^{(3)}$-decomposition of $K_{13}^{(3)} \setminus K_5^{(3)}$ consists of the orbits of the H-blocks in B under the action of the map $\infty \mapsto \infty$ and $j \mapsto j+1 \pmod{12}$.

Example 12 Let $V\left(K_{14}^{(3)} \setminus K_6^{(3)}\right) = \mathbb{Z}_{12} \cup \{\infty_1, \infty_2\}$ with $0, 3, 6, 9, \infty_1, \infty_2$ being the vertices in the hole and let

$$B_1 = \{H[0, 3, 8, 6, 4, 1, 2, 5, 9],\ H[5, 10, 9, 6, 8, 0, 2, 4, 7],$$
$$H[9, 10, 11, 3, 2, 8, 1, 0, 4],\ H[6, 0, 1, 3, 7, 9, 2, 4, 11],$$
$$H[0, 1, \infty_1, 2, 4, 8, \infty_2, 5, 10],\ H[0, 1, \infty_2, 2, 4, 8, \infty_1, 5, 10]\},$$

$$B_2 = \{H[1, 4, \infty_1, 2, \infty_2, 8, 5, 6, 10],\ H[2, 5, \infty_1, 4, \infty_2, 10, 7, 6, 11],$$
$$H[4, 7, \infty_1, 5, \infty_2, 11, 8, 9, 1],\ H[5, 8, \infty_1, 7, \infty_2, 1, 10, 2, 9],$$
$$H[7, 10, \infty_1, 8, \infty_2, 2, 11, 0, 4],\ H[8, 11, \infty_1, 10, \infty_2, 4, 1, 5, 0],$$
$$H[10, 1, \infty_1, 11, \infty_2, 5, 2, 3, 7],\ H[11, 2, \infty_1, 1, \infty_2, 7, 4, 8, 3],$$
$$H[0, 4, 8, \infty_2, 2, 6, 1, 7, \infty_1],\ H[1, 7, 4, \infty_1, 10, 3, 11, 5, \infty_2],$$
$$H[5, 9, 1, \infty_2, 7, 0, 8, 11, 2],\ H[1, 4, 10, 6, 2, 8, 5, 11, \infty_1],$$
$$H[7, 10, 4, 9, 5, 11, 8, \infty_1, 2],\ H[\infty_2, 4, 10, 1, 7, 3, 11, 2, 5]\}.$$

Then an $LP_4^{(3)}$-decomposition of $K_{14}^{(3)} \setminus K_6^{(3)}$ consists of the orbits of the H-blocks in B_1 under the action of the map $\infty_i \mapsto \infty_i$, for $i \in \{1, 2\}$, and $j \mapsto j + 1 \pmod{12}$ along with the H-blocks in B_2.

Example 13 Let $V\left(K_{15}^{(3)} \setminus K_7^{(3)}\right) = \mathbb{Z}_{12} \cup \{\infty_1, \infty_2, \infty_3\}$, with $0, 3, 6, 9, \infty_1, \infty_2, \infty_3$ being the vertices in the hole and let

$$B_1 = \{H[0, 3, 8, 6, 4, 1, 2, 5, 9],\ H[5, 10, 9, 6, 8, 0, 2, 4, 7],$$
$$H[9, 10, 11, 3, 2, 8, 1, 0, 4],\ H[6, 0, 1, 3, 7, 9, 2, 4, 11],$$
$$H[0, 1, \infty_1, 2, 4, 8, \infty_2, 5, 10],\ H[0, 1, \infty_2, 2, 4, 8, \infty_3, 5, 10],$$
$$H[0, 1, \infty_3, 2, 4, 8, \infty_1, 5, 10]\},$$

$$B_2 = \{H[4, 7, \infty_1, 1, \infty_2, 2, 5, 8, 11],\ H[\infty_1, \infty_3, 11, 8, \infty_2, 7, 4, 10, 1],$$
$$H[3, 4, 8, 5, \infty_2, 10, 7, 6, 11],\ H[\infty_2, 1, 7, \infty_3, 10, 3, 11, 2, 5],$$
$$H[2, 11, \infty_1, 8, \infty_2, 10, 1, 5, 9],\ H[1, 4, \infty_1, 10, \infty_2, 11, 2, 5, 8],$$
$$H[2, 5, \infty_1, 11, \infty_2, 1, 4, 7, 10],\ H[4, 10, \infty_1, 1, \infty_3, 8, 2, 3, 7],$$
$$H[1, 6, 2, \infty_1, \infty_3, 8, 11, 0, 4],\ H[5, 11, \infty_1, 4, \infty_3, 1, 7, 8, 0],$$
$$H[1, 7, \infty_1, 5, \infty_3, 4, 10, 2, 9],\ H[2, 8, \infty_2, 1, \infty_3, 7, 4, 5, 9],$$
$$H[4, 10, \infty_2, 2, \infty_3, 5, 8, 9, 1],\ H[6, 10, 2, 8, \infty_1, 7, \infty_3, 5, 11],$$
$$H[0, 4, 8, \infty_1, \infty_3, \infty_2, 10, 7, 1],\ H[5, 6, 10, \infty_1, \infty_3, \infty_2, 11, 3, 7],$$
$$H[8, 2, 11, \infty_2, 5, 0, 1, 4, 7],\ H[5, 8, \infty_1, 2, \infty_2, 4, \infty_3, 10, 1],$$

$$H[7, 10, \infty_1, 4, \infty_2, 5, \infty_3, 11, 2],\ H[8, 11, \infty_1, 5, \infty_2, 7, \infty_3, 4, 1],$$
$$H[1, 10, \infty_1, 7, \infty_2, 8, \infty_3, 2, 5]\big\}.$$

Then an $LP_4^{(3)}$-decomposition of $K_{15}^{(3)} \setminus K_7^{(3)}$ consists of the orbits of the H-blocks in B_1 under the action of the map $\infty_i \mapsto \infty_i$, for $i \in \{1, 2, 3\}$, and $j \mapsto j + 1 \pmod{12}$ along with the H-blocks in B_2.

Example 14 Let $V\left(K_{16}^{(3)} \setminus K_8^{(3)}\right) = \mathbb{Z}_8 \cup \{\infty_1, \infty_2, \ldots, \infty_8\}$, with $\infty_1, \infty_2, \ldots, \infty_8$ being the vertices in the hole and let

$$B_1 = \big\{H[0, \infty_1, \infty_2, 1, \infty_3, 2, \infty_4, 3, \infty_5],\ H[0, \infty_1, 1, 5, 3, \infty_2, 6, 4, \infty_3],$$
$$H[7, \infty_7, 6, \infty_8, 3, \infty_1, 0, \infty_5, 1],\ H[\infty_2, \infty_5, 0, \infty_1, \infty_3, \infty_6, 1, \infty_4, \infty_7],$$
$$H[6, \infty_8, \infty_6, \infty_2, 2, \infty_7, 0, \infty_4, \infty_1],\ H[7, \infty_7, \infty_2, 0, 2, 1, \infty_8, 5, \infty_5],$$
$$H[\infty_1, \infty_6, 4, \infty_4, 1, 0, \infty_3, 6, \infty_5],\ H[0, \infty_3, 3, \infty_6, 4, \infty_5, 6, \infty_4, \infty_8],$$
$$H[3, \infty_6, 0, \infty_4, 6, \infty_3, \infty_7, \infty_5, 7],\ H[\infty_1, \infty_5, 0, \infty_7, 3, 5, \infty_8, 6, \infty_2],$$
$$H[0, \infty_3, \infty_8, \infty_7, 7, \infty_2, \infty_4, 5, \infty_6],\ H[5, \infty_8, \infty_1, 2, \infty_7, 1, \infty_6, 0, \infty_5],$$
$$H[0, 1, 2, \infty_5, 5, \infty_6, 7, 6, \infty_4]\big\},$$

$$B_2 = \big\{H[\infty_1, 6, 2, 5, 1, 3, 0, 4, \infty_2],\ H[\infty_1, 7, 3, 6, 2, 4, 1, 5, \infty_2],$$
$$H[\infty_1, 0, 4, 7, 3, 5, 2, 6, \infty_2],\ H[\infty_1, 1, 5, 0, 4, 6, 3, 7, \infty_2],$$
$$H[\infty_3, 2, 6, 1, 5, 7, 4, 0, \infty_4],\ H[\infty_3, 3, 7, 2, 6, 0, 5, 1, \infty_4],$$
$$H[\infty_3, 4, 0, 3, 7, 1, 6, 2, \infty_4],\ H[\infty_3, 5, 1, 4, 0, 2, 7, 3, \infty_4],$$
$$H[\infty_5, 0, 4, 1, 5, 6, 3, 7, \infty_6],\ H[\infty_5, 1, 5, 2, 6, 7, 4, 0, \infty_6],$$
$$H[\infty_5, 2, 6, 3, 7, 0, 5, 1, \infty_6],\ H[\infty_5, 3, 7, 4, 0, 1, 6, 2, \infty_6],$$
$$H[\infty_7, 4, 0, 5, 1, 2, 7, 3, \infty_8],\ H[\infty_7, 5, 1, 6, 2, 3, 0, 4, \infty_8],$$
$$H[\infty_7, 6, 2, 7, 3, 4, 1, 5, \infty_8],\ H[\infty_7, 7, 3, 0, 4, 5, 2, 6, \infty_8],$$
$$H[0, 2, \infty_1, 1, 3, 4, \infty_2, 5, 6],\ H[2, 4, \infty_1, 5, 3, 1, 6, 7, \infty_2],$$
$$H[4, 6, \infty_1, 5, 7, 0, \infty_2, 1, 2],\ H[0, 2, 5, \infty_2, 4, 6, 1, 7, \infty_1],$$
$$H[4, 7, 2, \infty_2, 3, 5, 0, 6, \infty_1],\ H[3, 6, 0, \infty_2, 1, 4, 7, 5, 2]\big\}.$$

Then an $LP_4^{(3)}$-decomposition of $K_{16}^{(3)} \setminus K_8^{(3)}$ consists of the orbits of the H-blocks in B_1 under the action of the map $\infty_i \mapsto \infty_i$, for $i \in \{1, 2, \ldots, 8\}$, and $j \mapsto j + 1 \pmod{8}$ along with the H-blocks in B_2.

Example 15 Let $V\left(K_{11}^{(3)}\right) = \mathbb{Z}_{10} \cup \{\infty\}$ and let

13. G.B. Khosrovshahi, R. Laue, t-designs with $t \geq 3$. *Handbook of Combinatorial Designs*, ed. by C.J. Colbourn, J.H. Dinitz, 2nd edn. (Chapman & Hall/CRC Press, Boca Raton, 2007), pp. 79–101
14. J. Kuhl, M.W. Schroeder, Hamilton cycle decompositions of k-uniform k-partite hypergraphs. Australas. J. Combin. **56**, 23–37 (2013)
15. D. Kühn, D. Osthus, Decompositions of complete uniform hypergraphs into Hamilton Berge cycles. J. Combin. Theory Ser. A **126**, 128–135 (2014)
16. Z. Lonc, Solution of a delta-system decomposition problem. J. Combin. Theory, Ser. A **55**, 33–48 (1990)
17. Z. Lonc, Packing, covering and decomposing of a complete uniform hypergraph into delta-systems. Graphs Combin. **8**, 333–341(1992)
18. M. Meszka, A. Rosa, Decomposing complete 3-uniform hypergraphs into Hamiltonian cycles. Australas. J. Combin. **45**, 291–302 (2009)
19. A.-F. Mouyart, F. Sterboul, Decomposition of the complete hypergraph into delta-systems II. J. Combin. Theory, Ser. A **41**, 139–149 (1986)
20. M.W. Schroeder, On Hamilton cycle decompositions of r-uniform r-partite hypergraphs. Discrete Math. **315**, 1–8 (2014)
21. R.M. Wilson, Decompositions of Complete Graphs into Subgraphs Isomorphic to a Given Graph, in *Proceedings of the Fifth British Combinatorial Conference*, ed. by C.S.J.A. Nash-Williams, J. Sheehan. Congressus Numerantium, vol. XV (1975), pp. 647–659

On Loose 5-Cycle Decompositions, Packings, and Coverings of λ-Fold Complete 3-Uniform Hypergraphs

Ryan C. Bunge, Joshua Jetton, Marissa Juarez, Anthony Netz, Dan Roberts, and Patrick Ward

Abstract The λ-fold complete 3-uniform hypergraph of order v, denoted ${}^{\lambda}K_v^{(3)}$, is the hypergraph on v vertices with edge multiset consisting of λ copies of each of the 3-element subsets of the vertex set. A loose 5-cycle, denoted LC_5, is a hypergraph with vertex set $\{v_1, v_2, v_3, v_4, v_5, v_6, v_7, v_8, v_9, v_{10}\}$ and edge set $\big\{\{v_1, v_2, v_3\}, \{v_3, v_4, v_5\}, \{v_5, v_6, v_7\}, \{v_7, v_8, v_9\}, \{v_9, v_{10}, v_1\}\big\}$. We give necessary and sufficient conditions for the existence of an LC_5-decomposition of ${}^{\lambda}K_v^{(3)}$. We also characterize the number of edges in the leave of a maximum LC_5-packing and the padding of a minimum LC_5-covering of ${}^{\lambda}K_v^{(3)}$.

Keywords 3-uniform hypergraph · Hypergraph decomposition · Maximum packing · Minimum covering · Loose 5-cycle

MSC Codes 05C51, 05B40, 05C65

R. C. Bunge (✉)
Illinois State University, Normal, IL, USA
e-mail: rcbunge@ilstu.edu

J. Jetton
Eastern Michigan University, Ypsilanti, MI, USA

M. Juarez
Dominican University, River Forest, IL, USA

A. Netz
Oregon State University, Corvallis, OR, USA

D. Roberts · P. Ward
Illinois Wesleyan University, Bloomington, IL, USA
e-mail: drobert1@iwu.edu

A. Bahmanian (ed.), *Combinatorics and Combinatorial Computing*, Springer Proceedings in Mathematics & Statistics 516,
https://doi.org/10.1007/978-3-032-08549-8_5

1 Introduction

A commonly studied problem in combinatorics concerns decompositions of graphs into edge-disjoint subgraphs. A *decomposition* of a graph K is a set $\Delta = \{G_1, G_2, \ldots, G_s\}$ of pairwise edge-disjoint subgraphs of K such that $E(G_1) \cup E(G_2) \cup \cdots \cup E(G_s) = E(K)$. If each element of Δ is isomorphic to a fixed graph G, then Δ is called a *G-decomposition* of K. A G-decomposition of K_v is also known as a *G-design of order v*. A K_k-design of order v is an *$S(2, k, v)$-design* or a *Steiner system*. An $S(2, k, v)$-design is also known as a *balanced incomplete block design of index 1* or a *$(v, k, 1)$-BIBD*. The problem of determining all v for which there exists a G-design of order v is of special interest (see [1] for a survey).

The notion of decompositions of graphs naturally extends to decompositions of uniform hypergraphs. A *hypergraph* H consists of a finite nonempty set V of *vertices* and a set $E = \{e_1, e_2, \ldots, e_m\}$ of nonempty subsets of V called *edges* (or sometimes, *hyperedges*). If E is allowed to be a multiset, then H is called a *multi-hypergraph*. If E is not a multiset and does not contain any multisets, then H is called *simple*. If for each $e \in E$ we have $|e| = t$, then H is said to be *t-uniform*. Thus loopless graphs are 2-uniform hypergraphs. The complete t-uniform hypergraph on the vertex set V has the set of all t-element subsets of V as its edge set and is denoted by $K_V^{(t)}$. The symbol $K_v^{(t)}$ denotes the *complete t-uniform hypergraph of order v*, that is, any hypergraph isomorphic to $K_V^{(t)}$ with $|V| = v$. Graphs with repeated edges are often referred to as *multigraphs* to emphasize that edges are repeated. If H is a simple hypergraph and λ is a positive integer, then *λ-fold H*, denoted ${}^{\lambda}H$, is the multi-hypergraph obtained from H by repeating each edge exactly λ times. In the context of this paper, a *subgraph* of a hypergraph, H, is a hypergraph, H', such that $V(H') \subseteq V(H)$ and $E(H') \subseteq E(H)$.

A *decomposition* of a hypergraph K is a set $\Delta = \{H_1, H_2, \ldots, H_s\}$ of pairwise edge-disjoint subgraphs of K such that $E(H_1) \cup E(H_2) \cup \cdots \cup E(H_s) = E(K)$. If each element H_i of Δ is isomorphic to a fixed hypergraph H, then each H_i is called an *H-block*, and Δ is called an *H-decomposition* of K. If there exists an H-decomposition of K, then we may simply state that *H decomposes K*. Assuming H is t-uniform, an H-decomposition of the (λ-fold) complete t-uniform hypergraph of order v is also called an *H-design of order v (and index λ)*. The problem of determining all v for which there exists an H-design of order v is called the *spectrum problem for H-designs*.

When an H-decomposition does not exist, it is natural to ask how close one can get to such a structure. To this end we define two structures that approximate decompositions: maximum packings and minimum coverings. Let K be a hypergraph. An *H-packing of K* is a collection of edge-disjoint subgraphs of K, each of which is isomorphic to H. The collection of edges of K that are not present in any of the copies of H is called the *leave* of the H-packing. An H-packing is called a *maximum H-packing* when the cardinality of the leave is minimum with respect to all H-packings of K. Similarly, an *H-covering of K* is a collection of copies of H which use every edge in K at least once. The multiset difference of the multiset union of the edge sets of all of these copies of H and the edge set of K is called the *padding* of the H-covering. An H-covering is called a *minimum H-covering* when the cardinality of the padding is minimum with respect to all H-coverings of K.

A $K_k^{(t)}$-design of order v is a generalization of Steiner systems and is equivalent to an $S(t, k, v)$-design. A summary of results on $S(t, k, v)$-designs appears in [10]. Keevash [15] has recently shown that for all t and k the obvious necessary conditions for the existence of an $S(t, k, v)$-design are sufficient for sufficiently large values of v. Similar results were obtained by Glock, Kühn, Lo, and Osthus [11] and extended to include the corresponding asymptotic results for H-designs of order v for all uniform hypergraphs H. These results for t-uniform hypergraphs mirror the celebrated results of Wilson [21] for graphs. Although these asymptotic results assure the existence of H-designs for sufficiently large values of v for any uniform hypergraph H, the spectrum problem has been settled for very few t-uniform hypergraphs H with $t > 2$.

In the study of graph decompositions, a fair amount of the focus has been on G-decompositions of K_v where G is a graph with a relatively small number of edges (see [1] and [5] for known results). Some authors have investigated the corresponding problem for 3-uniform hypergraphs. For example, in [4], the spectrum problem is settled for all 3-uniform hypergraphs on 4 or fewer vertices. More recently, the spectrum problem was settled in [6] for all 3-uniform hypergraphs with at most 6 vertices and at most 3 edges. In [6], the authors also settle the spectrum problem for the 3-uniform hypergraph of order 6 whose edges form the lines of the Pasch configuration. Authors have also considered H-designs where H is a 3-uniform hypergraph whose edge set is determined by the faces of a regular polyhedron. Let T, O, and I denote the tetrahedron, the octahedron, and the icosahedron hypergraphs, respectively. The hypergraph T is the same as $K_4^{(3)}$, and its spectrum was settled in 1960 by Hanani [12]. In another paper [13], Hanani settled the spectrum problem for O-designs and gave necessary conditions for the existence of I-designs. Perhaps the best known general result on decompositions of complete t-uniform hypergraphs is Baranyai's result [3] on the existence of 1-factorizations of $K_{mt}^{(t)}$ for all positive integers m. There are, however, several articles on decompositions of complete t-uniform hypergraphs (see [2] and [18]) and of t-uniform t-partite hypergraphs (see [16] and [20]) into variations on the concept of a Hamilton cycle. There are also several results on decompositions of 3-uniform hypergraphs into structures known as Berge cycles with a given number of edges (see for example [14] and [17]). We note however that the Berge cycles in these decompositions are not required to be isomorphic. Berge cycles allow for any number of vertices to be contained in the intersection of consecutive edges, whereas loose cycles specify that these intersections contain exactly one vertex.

In this paper we are interested in the spectrum problem for H-designs where H is the hypergraph known as a loose 5-cycle. A *loose m-cycle* in $K_n^{(3)}$, denoted LC_m, is a hypergraph with vertex set $\{v_1, v_2, \ldots, v_{2m}\}$ and edge set $\big\{\{v_{2i-1}, v_{2i}, v_{2i+1}\} : 1 \leq i \leq m-1\big\} \cup \big\{\{v_{2m-1}, v_{2m}, v_1\}\big\}$. Let $H[v_1, v_2, v_3, v_4, v_5, v_6, v_7, v_8, v_9, v_{10}]$ denote the loose 5-cycle with vertex set $\{v_1, v_2, v_3, v_4, v_5, v_6, v_7, v_8, v_9, v_{10}\}$ and edge set

Let Δ be the union of the orbits of the cycles in B under the action of the map $j \mapsto j+1 \pmod{11}$ and where ∞ is fixed by the mapping. It can be verified that Δ is an LC_5-decomposition of $K_{12}^{(3)}$.

Example 4 Let $V\left(K_{15}^{(3)}\right) = \mathbb{Z}_{13} \cup \{\infty_1, \infty_2\}$ and let

$$B = \{H[0,6,1,4,11,2,8,10,9,7],\ H[0,1,\infty_2,2,5,9,6,7,\infty_1,10],\ H[0,2,8,6,\infty_2,7,12,10,\infty_1,5],\ H[0,7,2,1,10,11,5,6,3,9],\ H[0,6,7,11,\infty_2,9,\infty_1,8,4,2],\ H[0,3,8,6,5,1,10,2,12,4],\ H[0,\infty_2,7,4,9,11,5,2,6,\infty_1]\}.$$

Let Δ be the union of the orbits of the cycles in B under the action of the map $j \mapsto j+1 \pmod{13}$ and where $\infty_i \mapsto \infty_i$ for $i \in \{1,2\}$. It can be verified that Δ is an LC_5-decomposition of $K_{15}^{(3)}$.

Example 5 Let $V\left(K_{16}^{(3)}\right) = \mathbb{Z}_{16}$ and let

$$B = \{H[0,4,1,5,15,10,11,6,14,12],\ H[0,9,2,14,7,13,8,10,3,6],\ H[0,2,1,13,3,12,6,10,14,5],\ H[0,8,1,13,4,14,9,15,5,6],\ H[0,5,2,1,8,15,4,13,3,7],\ H[0,3,1,2,14,15,12,4,7,8],\ H[0,3,10,9,5,7,13,11,8,6]\}.$$

Let Δ be the union of the orbits of the cycles in B under the action of the map $j \mapsto j+1 \pmod{16}$. It can be verified that Δ is an LC_5-decomposition of $K_{16}^{(3)}$.

Example 6 Let $V\left(K_{17}^{(3)}\right) = \mathbb{Z}_{17}$ and let

$$B = \{H[0,5,2,10,6,7,14,3,1,4],\ H[0,8,2,11,6,10,5,15,1,14],\ H[0,9,2,3,4,11,15,6,1,12],\ H[0,10,2,13,6,14,15,1,7,12],\ H[0,6,3,9,5,10,16,11,1,7],\ H[0,8,14,11,16,4,3,5,1,15],\ H[0,12,3,10,5,4,15,11,1,6],\ H[0,11,2,14,6,5,8,15,1,10]\}.$$

Let Δ be the union of the orbits of the cycles in B under the action of the map $j \mapsto j+1 \pmod{17}$. It can be verified that Δ is an LC_5-decomposition of $K_{17}^{(3)}$.

Example 7 Let $V\left(L_{5,5}^{(3)}\right) = \mathbb{Z}_{10}$ with vertex set partition $\{\{0,2,4,6,8\},\{1,3,5,7,9\}\}$ and let

$$B = \{H[0,3,6,7,8,1,9,4,2,5],\ H[0,7,1,5,2,3,8,6,9,4]\}.$$

Let Δ be the union of the orbits of the cycles in B under the action of the map $j \mapsto j+1 \pmod{10}$. It can be verified that Δ is an LC_5-decomposition of $L^{(3)}_{5,5}$.

Example 8 Let $V\left(L^{(3)}_{5,5} \cup K^{(3)}_{1,5,5}\right) = \mathbb{Z}_{10} \cup \{\infty\}$ with vertex set partition $\{\{\infty\}, \{0, 2, 4, 6, 8\}, \{1, 3, 5, 7, 9\}\}$ and let

$$B_1 = \{H[0, 1, \infty, 3, 6, 2, 9, 5, 8, 7],\ H[0, 1, 5, 4, 8, 3, 6, 2, 7, 9]\},$$

$$B_2 = \{H[0, \infty, 5, 3, 8, 9, 7, 4, 2, 1],\ H[1, \infty, 6, 7, 8, 0, 3, 5, 4, 9],$$
$$H[9, \infty, 4, 5, 6, 8, 1, 3, 2, 7],\ H[7, \infty, 2, 3, 4, 6, 9, 1, 0, 5],$$
$$H[8, \infty, 3, 1, 6, 7, 5, 2, 0, 9]\}.$$

Let Δ be the union of B_2 and the orbits of the cycles in B_1 under the action of the map $j \mapsto j+1 \pmod{10}$ and $\infty \mapsto \infty$. It can be verified that Δ is an LC_5-decomposition of $L^{(3)}_{5,5} \cup K^{(3)}_{1,5,5}$.

Example 9 Let $V\left(L^{(3)}_{5,5} \cup K^{(3)}_{2,5,5}\right) = \mathbb{Z}_{10} \cup \{\infty_1, \infty_2\}$ with vertex set partition $\{\{\infty_1, \infty_2\}, \{0, 2, 4, 6, 8\}, \{1, 3, 5, 7, 9\}\}$ and let

$$B_1 = \{H[0, 3, 6, \infty_2, 7, 5, 8, 4, 9, \infty_1],\ H[0, 1, 2, \infty_1, 5, 8, 4, 9, 7, \infty_2]\},$$

$$B_2 = \{H[0, 3, 1, 2, 6, 4, 9, 8, 5, \infty_1],\ H[1, 4, 2, 3, 7, 5, 0, 9, 6, \infty_1],$$
$$H[2, 5, 3, 4, 8, 6, 1, 0, 7, \infty_1],\ H[3, 6, 4, 5, 9, 7, 2, 1, 8, \infty_1],$$
$$H[4, 7, 5, 6, 0, 8, 3, 2, 9, \infty_1],\ H[0, 3, 4, 9, 1, 7, 6, 8, 5, \infty_2],$$
$$H[1, 4, 5, 0, 2, 8, 7, 9, 6, \infty_2],\ H[2, 5, 6, 1, 3, 9, 8, 0, 7, \infty_2],$$
$$H[3, 6, 7, 2, 4, 0, 9, 1, 8, \infty_2],\ H[4, 7, 8, 3, 5, 1, 0, 2, 9, \infty_2]\}.$$

Let Δ be the union of B_2 and the orbits of the cycles in B_1 under the action of the map $j \mapsto j+1 \pmod{10}$ and where $\infty_i \mapsto \infty_i$ for $i \in \{1, 2\}$. It can be verified that Δ is an LC_5-decomposition of $L^{(3)}_{5,5} \cup K^{(3)}_{2,5,5}$.

Example 10 Let $V\left(K^{(3)}_{3,5,5}\right) = \mathbb{Z}_{10} \cup \{\infty_1, \infty_2, \infty_3\}$ with vertex set partition $\{\{\infty_1, \infty_2, \infty_3\}, \{0, 2, 4, 6, 8\}, \{1, 3, 5, 7, 9\}\}$ and let

$$B = \{H[0, \infty_1, 5, 2, \infty_3, 7, 6, 3, \infty_2, 1],\ H[1, \infty_1, 6, 3, \infty_3, 8, 7, 4, \infty_2, 2],$$
$$H[2, \infty_1, 7, 4, \infty_3, 9, 8, 5, \infty_2, 3],\ H[3, \infty_1, 8, 5, \infty_3, 0, 9, 6, \infty_2, 4],$$
$$H[4, \infty_1, 9, 6, \infty_3, 1, 0, 7, \infty_2, 5],\ H[0, \infty_2, 5, 8, \infty_1, 3, 2, 1, \infty_3, 7],$$
$$H[1, \infty_2, 6, 9, \infty_1, 4, 3, 2, \infty_3, 8],\ H[2, \infty_2, 7, 0, \infty_1, 5, 4, 3, \infty_3, 9],$$
$$H[3, \infty_2, 8, 1, \infty_1, 6, 5, 4, \infty_3, 0],\ H[4, \infty_2, 9, 2, \infty_1, 7, 6, 5, \infty_3, 1],$$

$$H[0, \infty_3, 5, 6, \infty_2, 1, 8, 7, \infty_1, 3],\ H[1, \infty_3, 6, 7, \infty_2, 2, 9, 8, \infty_1, 4],$$
$$H[2, \infty_3, 7, 8, \infty_2, 3, 0, 9, \infty_1, 5],\ H[3, \infty_3, 8, 9, \infty_2, 4, 1, 0, \infty_1, 6],$$
$$H[4, \infty_3, 9, 0, \infty_2, 5, 2, 1, \infty_1, 7]\big\}.$$

Then B is an LC_5-decomposition of $K^{(3)}_{3,5,5}$.

Example 11 Let $V\left(K^{(3)}_{4,5,5}\right) = \mathbb{Z}_{10} \cup \{\infty_1, \infty_2, \infty_3, \infty_4\}$ with vertex set partition $\big\{\{\infty_1, \infty_2, \infty_3, \infty_4\}, \{0, 2, 4, 6, 8\}, \{1, 3, 5, 7, 9\}\big\}$ and let

$$B = \big\{H[0, 1, \infty_3, 6, 3, 4, \infty_4, 2, 5, \infty_1],\ H[1, 2, \infty_3, 7, 4, 5, \infty_4, 3, 6, \infty_1],$$
$$H[2, 3, \infty_3, 8, 5, 6, \infty_4, 4, 7, \infty_1],\ H[3, 4, \infty_3, 9, 6, 7, \infty_4, 5, 8, \infty_1],$$
$$H[4, 5, \infty_3, 0, 7, 8, \infty_4, 6, 9, \infty_1],\ H[0, \infty_2, 5, 6, \infty_3, 1, 8, 9, \infty_4, 7],$$
$$H[1, \infty_2, 6, 7, \infty_3, 2, 9, 0, \infty_4, 8],\ H[2, \infty_2, 7, 8, \infty_3, 3, 0, 1, \infty_4, 9],$$
$$H[3, \infty_2, 8, 9, \infty_3, 4, 1, 2, \infty_4, 0],\ H[4, \infty_2, 9, 0, \infty_3, 5, 2, 3, \infty_4, 1],$$
$$H[0, 1, \infty_1, 6, 3, 4, \infty_2, 2, 5, \infty_3],\ H[1, 2, \infty_1, 7, 4, 5, \infty_2, 3, 6, \infty_3],$$
$$H[2, 3, \infty_1, 8, 5, 6, \infty_2, 4, 7, \infty_3],\ H[3, 4, \infty_1, 9, 6, 7, \infty_2, 5, 8, \infty_3],$$
$$H[4, 5, \infty_1, 0, 7, 8, \infty_2, 6, 9, \infty_3],\ H[0, \infty_4, 5, 6, \infty_1, 1, 8, 9, \infty_2, 7],$$
$$H[1, \infty_4, 6, 7, \infty_1, 2, 9, 0, \infty_2, 8],\ H[2, \infty_4, 7, 8, \infty_1, 3, 0, 1, \infty_2, 9],$$
$$H[3, \infty_4, 8, 9, \infty_1, 4, 1, 2, \infty_2, 0],\ H[4, \infty_4, 9, 0, \infty_1, 5, 2, 3, \infty_2, 1]\big\}.$$

Then B is an LC_5-decomposition of $K^{(3)}_{4,5,5}$.

Example 12 Let $V\left(K^{(3)}_{5,5,5}\right) = \mathbb{Z}_{15}$ with vertex set partition $\big\{\{0, 3, 6, 9, 12\}, \{1, 4, 7, 10, 13\}, \{2, 5, 8, 11, 14\}\big\}$ and let

$$B = \big\{H[0, 14, 1, 11, 3, 10, 5, 6, 4, 2],\ H[0, 2, 1, 5, 6, 11, 10, 12, 8, 7],$$
$$H[0, 11, 1, 8, 9, 13, 2, 12, 10, 5],\ H[0, 5, 1, 12, 8, 10, 3, 14, 13, 2],$$
$$H[0, 1, 8, 4, 12, 7, 5, 6, 10, 2]\big\}.$$

Let Δ be the union of the orbits of the cycles in B under the action of the map $j \mapsto j + 3 \pmod{15}$. It can be verified that Δ is an LC_5-decomposition of $K^{(3)}_{5,5,5}$.

Example 13 Let $V\left(K^{(3)}_{13} \setminus K^{(3)}_{3}\right) = \mathbb{Z}_{10} \cup \{\infty_1, \infty_2, \infty_3\}$ with $\infty_1, \infty_2, \infty_3$ being the vertices in the hole and let

$$B_1 = \{H[0,6,2,8,3,5,\infty_1,4,\infty_2,1],\ H[0,6,3,7,4,1,\infty_1,5,\infty_3,2],$$
$$H[0,6,9,4,7,1,\infty_2,2,\infty_3,3],\ H[0,\infty_1,4,\infty_3,8,\infty_2,1,9,2,5]\},$$

$$B_2 = \{H[0,5,1,6,2,7,3,8,4,9],\ H[5,0,6,1,7,2,8,3,9,4],$$
$$H[0,2,1,3,\infty_2,8,6,7,5,\infty_1],\ H[1,3,2,4,\infty_2,9,7,8,6,\infty_1],$$
$$H[2,4,3,5,\infty_2,0,8,9,7,\infty_1],\ H[3,5,4,6,\infty_2,1,9,0,8,\infty_1],$$
$$H[4,6,5,7,\infty_2,2,0,1,9,\infty_1],\ H[0,3,1,2,\infty_3,7,6,8,5,\infty_2],$$
$$H[1,4,2,3,\infty_3,8,7,9,6,\infty_2],\ H[2,5,3,4,\infty_3,9,8,0,7,\infty_2],$$
$$H[3,6,4,5,\infty_3,0,9,1,8,\infty_2],\ H[4,7,5,6,\infty_3,1,0,2,9,\infty_2],$$
$$H[0,4,2,3,\infty_1,8,7,9,5,\infty_3],\ H[1,5,3,4,\infty_1,9,8,0,6,\infty_3],$$
$$H[2,6,4,5,\infty_1,0,9,1,7,\infty_3],\ H[3,7,5,6,\infty_1,1,0,2,8,\infty_3],$$
$$H[4,8,6,7,\infty_1,2,1,3,9,\infty_3]\}.$$

Let Δ be the union of B_2 and the orbits of the cycles in B_1 under the action of the map $j \mapsto j+1 \pmod{10}$ and where $\infty_i \mapsto \infty_i$ for $i \in \{1,2,3\}$. It can be verified that Δ is an LC_5-decomposition of $K_{13}^{(3)} \setminus K_3^{(3)}$.

Example 14 Let $V\left(K_{14}^{(3)} \setminus K_4^{(3)}\right) = \mathbb{Z}_{10} \cup \{\infty_1,\infty_2,\infty_3,\infty_4\}$ with $\infty_1,\infty_2,\infty_3,\infty_4$ being the vertices in the hole and let

$$B_1 = \{H[0,3,6,1,4,2,\infty_2,8,\infty_1,7],\ H[0,6,2,5,\infty_4,3,\infty_1,1,\infty_3,4],$$
$$H[0,5,9,6,8,4,\infty_4,2,\infty_3,1],\ H[0,\infty_3,7,3,6,4,\infty_4,1,\infty_2,3],$$
$$H[0,4,\infty_2,6,\infty_3,3,1,\infty_4,2,5]\},$$

$$B_2 = \{H[0,5,1,6,2,7,3,8,4,9],\ H[5,0,6,1,7,2,8,3,9,4],$$
$$H[0,3,1,2,\infty_2,7,6,8,5,\infty_1],\ H[1,4,2,3,\infty_2,8,7,9,6,\infty_1],$$
$$H[2,5,3,4,\infty_2,9,8,0,7,\infty_1],\ H[3,6,4,5,\infty_2,0,9,1,8,\infty_1],$$
$$H[4,7,5,6,\infty_2,1,0,2,9,\infty_1],\ H[0,4,1,2,\infty_1,7,6,9,5,\infty_3],$$
$$H[1,5,2,3,\infty_1,8,7,0,6,\infty_3],\ H[2,6,3,4,\infty_1,9,8,1,7,\infty_3],$$
$$H[3,7,4,5,\infty_1,0,9,2,8,\infty_3],\ H[4,8,5,6,\infty_1,1,0,3,9,\infty_3],$$
$$H[0,2,1,3,\infty_1,8,6,7,5,\infty_2],\ H[1,3,2,4,\infty_1,9,7,8,6,\infty_2],$$
$$H[2,4,3,5,\infty_1,0,8,9,7,\infty_2],\ H[3,5,4,6,\infty_1,1,9,0,8,\infty_2],$$
$$H[4,6,5,7,\infty_1,2,0,1,9,\infty_2],\ H[0,4,2,6,\infty_1,1,7,9,5,\infty_4],$$
$$H[1,5,3,7,\infty_1,2,8,0,6,\infty_4],\ H[2,6,4,8,\infty_1,3,9,1,7,\infty_4],$$
$$H[3,7,5,9,\infty_1,4,0,2,8,\infty_4],\ H[4,8,6,0,\infty_1,5,1,3,9,\infty_4]\}.$$

Lemma 1 *There exists an* LC_5*-decomposition of* $K_v^{(3)}$ *only if* $v \equiv 0, 1,$ *or* $2 \pmod 5$ *and* $v \notin \{5, 6, 7\}$.

We show that the above conditions are sufficient by constructing LC_5-decompositions of $K_v^{(3)}$ for all $v \equiv 0, 1,$ or $2 \pmod 5$ with $v \geq 10$. Our constructions are dependent on the many small examples given in Sect. 2.

We begin by proving a lemma that is fundamental to our constructions.

Lemma 2 *For* $r \in \{0, 1, 2\}$ *and all positive integers* x *and* y*, there exists a decomposition of* $K_{r,5x,5y}^{(3)} \cup L_{5x,5y}^{(3)}$ *into copies of* $K_{5,5,5}^{(3)}$ *and* $K_{r,5,5}^{(3)} \cup L_{5,5}^{(3)}$.

Proof Let $r \in \{0, 1, 2\}$ and let x and y be positive integers. The vertices of $K_{r,5x,5y}^{(3)} \cup L_{5x,5y}^{(3)}$ can be partitioned into sets V_i, W_j, and R where $1 \leq i \leq x$, $1 \leq j \leq y$, $|V_i| = 5 = |W_j|$, and $|R| = r$ such that every edge $\{a, b, c\}$ is of exactly one of the following types:

Type 1: there exist i, j with $a \in R$, $b \in V_i$, and $c \in W_j$;

Type 2: there exist i, j, k with $i \neq j$, $a \in V_i$, $b \in V_j$, and $c \in W_k$;

Type 3: there exist i, j, k with $j \neq k$, $a \in V_i$, $b \in W_j$, and $c \in W_k$;

Type 4: there exist i, j with $a, b \in V_i$ and $c \in W_j$; or

Type 5: there exist i, j with $a \in V_i$ and $b, c \in W_j$.

For every choice of i and j we can put together the edges of Types 1, 4, and 5 to form a copy of $K_{r,5,5}^{(3)} \cup L_{5,5}^{(3)}$. For every choice of i, j, and k the edges of Types 2 and 3 form copies of $K_{5,5,5}^{(3)}$. Since all edges are accounted for by exactly one of the aforementioned choices of subscripts, we have the desired decomposition into copies of $K_{5,5,5}^{(3)}$ and $K_{r,5,5}^{(3)} \cup L_{5,5}^{(3)}$. □

Theorem 1 *There exists an* LC_5*-decomposition of* $K_v^{(3)}$ *if and only if* $v \equiv 0, 1,$ *or* $2 \pmod 5$ *and* $v \notin \{5, 6, 7\}$.

Proof The necessary conditions for the existence of an LC_5-decomposition of $K_v^{(3)}$ are established in Lemma 1. Thus we need only to establish their sufficiency. Let $v = 5x + r$ where $x \geq 2$ and $r \in \{0, 1, 2\}$. We next consider two cases depending on the parity of x.

When x is even we can write $K_v^{(3)}$ as the edge-disjoint union of copies of $K_{10+r}^{(3)}$, $K_{r,10,10}^{(3)} \cup L_{10,10}^{(3)}$, and $K_{10,10,10}^{(3)}$, where the group of r vertices is common to every applicable copy. By Examples 1, 2, and 3 we have that an LC_5-decomposition of $K_{10+r}^{(3)}$ exists. By Lemma 2 we have that $K_{r,10,10}^{(3)} \cup L_{10,10}^{(3)}$ can be decomposed into copies of $K_{5,5,5}^{(3)}$ and $K_{r,5,5}^{(3)} \cup L_{5,5}^{(3)}$. By Example 12 an LC_5-decomposition of $K_{5,5,5}^{(3)}$ exists. When $r = 0$ $K_{0,5,5}^{(3)} \cup L_{5,5}^{(3)}$ is isomorphic to $L_{5,5}^{(3)}$, which admits an LC_5-decomposition by Example 7. When $r \in \{1, 2\}$ an LC_5-decomposition of $K_{r,5,5}^{(3)} \cup L_{5,5}^{(3)}$ exists by Examples 8 and 9, respectively. Finally, it is straightforward to see that $K^{(3)}_{10,10,10}$ can be decomposed into copies of $K^{(3)}_{5,5,5}$; thus, by Example 12 an LC_5-decomposition of $K^{(3)}_{10,10,10}$ exists.

When x is odd the construction is similar, and we can write $K^{(3)}_v$ as the edge-disjoint union of copies of $K^{(3)}_{15+r}$, $K^{(3)}_{10+r}$, $K^{(3)}_{r,15,10} \cup L^{(3)}_{15,10}$, $K^{(3)}_{r,10,10} \cup L^{(3)}_{10,10}$, $K^{(3)}_{15,10,10}$, and $K^{(3)}_{10,10,10}$, where again the group of r vertices is common to every applicable copy. The existence of LC_5-decompositions of each of these hypergraphs can be established by using the ingredients listed in the previous case along with Examples 4, 5, and 6. □

Next, we give our main result on maximum packings.

Theorem 2 *Let $x \geq 2$ be an integer. The leave of a maximum LC_5-packing of $K^{(3)}_{5x+3}$ consists of a single edge. Furthermore, the leave of a maximum LC_5-packing of $K^{(3)}_{5x+4}$ consists of four edges, which can be configured as an LP_4.*

Proof Let $v = 5x + r$ where $x \geq 2$ and $r \in \{3, 4\}$. Using a similar argument as in the proof of Theorem 1, we consider two cases depending on the parity of x.

When x is even, we can write $K^{(3)}_v$ as the union of $K^{(3)}_{10+r}$ and copies of $K^{(3)}_{10+r} \setminus K^{(3)}_r$, $K^{(3)}_{r,10,10} \cup L^{(3)}_{10,10}$, and $K^{(3)}_{10,10,10}$. We take a maximum LC_5-packing on the $K^{(3)}_{10+r}$ with the desired leave, i.e., a single edge or an LP_4, which exists by Examples 13 and 16, respectively. On each of the copies of $K^{(3)}_{10+r} \setminus K^{(3)}_r$ we take an LC_5-decomposition of $K^{(3)}_{10+r} \setminus K^{(3)}_r$ where the hole is chosen to be the relevant copy of $K^{(3)}_r$. These exist by Examples 13 and 14. The hypergraph induced by the remaining edges in $K^{(3)}_v$ can be decomposed into copies of LC_5 in the same manner as detailed in the proof of Theorem 1.

When x is odd, the construction is similar: we can write $K^{(3)}_v$ as the union of $K^{(3)}_{15+r}$ and copies of $K^{(3)}_{10+r} \setminus K^{(3)}_r$, $K^{(3)}_{r,15,10} \cup L^{(3)}_{15,10}$, $K^{(3)}_{r,10,10} \cup L^{(3)}_{10,10}$, $K^{(3)}_{15,10,10}$, and $K^{(3)}_{10,10,10}$. We take a maximum LC_5-packing on the $K^{(3)}_{15+r}$ with the desired leave, i.e., a single edge or an LP_4, which exists by Examples 15 and 17, respectively. The hypergraph induced by the remainder of the edges admits an LC_5-decomposition as shown in Theorem 1.

Since the established LC_5-packing of $K^{(3)}_v$ has a leave with edges than an LC_5, this is a maximum packing. □

Theorem 3 *Let λ and v be positive integers with $v \geq 10$. Then an LC_5-decomposition of ${}^{\lambda}K^{(3)}_v$ exists if and only if*

(i) $\lambda \equiv 0 \pmod 5$ *or*
(ii) $v \equiv 0, 1,$ *or* $2 \pmod 5$.

Proof The necessity of these conditions follows from the requirement that the number of edges in ${}^{\lambda}K^{(3)}_v$ be a multiple of that in LC_5, i.e., 5 must divide $\lambda v(v-1)(v-2)/6$.

To establish sufficiency, we first note that if condition (ii) is satisfied, then an LC_5-decomposition of $K_v^{(3)}$ exists by Theorem 1. Putting together λ copies of this LC_5-decomposition produces the desired LC_5-decomposition of ${}^\lambda K_v^{(3)}$.

Now suppose that condition (i) is satisfied. Suppose $\lambda = 5x$ for some positive integer x. Notice that ${}^\lambda K_v^{(3)}$ can be decomposed into x copies of ${}^5 K_v^{(3)}$. Thus, if we can produce an LC_5-decomposition of ${}^5 K_v^{(3)}$, then we can put together x copies of it to produce the desired LC_5-decomposition of ${}^\lambda K_v^{(3)}$. When $v \equiv 0, 1, 2 \pmod 5$ such an LC_5-decomposition exists. So we must consider the remaining two cases.

When $v \equiv 3 \pmod 5$, a maximum LC_5-packing of $K_v^{(3)}$ has a leave of a single edge (by Theorem 2). Consider ${}^5 K_v^{(3)}$ to be the edge-disjoint union of five copies of $K_v^{(3)}$. Choose distinct vertices $v_1, v_2, \ldots, v_{10}$ of $K_v^{(3)}$ arbitrarily and place a maximum LC_5-packing on the five copies of $K_v^{(3)}$ in such a way that the leave edges in these five copies are, respectively, $\{v_1, v_2, v_3\}$, $\{v_3, v_4, v_5\}$, $\{v_5, v_6, v_7\}$, $\{v_7, v_8, v_9\}$, and $\{v_9, v_{10}, v_1\}$. These edges form one copy of LC_5; therefore, we have the desired LC_5-decomposition of $K_v^{(3)}$.

When $v \equiv 4 \pmod 5$, there exists a maximum LC_5-packing of $K_v^{(3)}$ whose leave is the edge set of an LP_4 (also by Theorem 2). Again, choose the distinct vertices $v_1, v_2, \ldots, v_{10}$ arbitrarily from $K_v^{(3)}$ and place a maximum LC_5-packing on each of those 5 copies of $K_v^{(3)}$ where the leave edges in these five copies are, respectively, the following:

Leave 1: $\{v_1, v_2, v_3\}$, $\{v_3, v_4, v_5\}$, $\{v_5, v_6, v_7\}$, $\{v_7, v_8, v_9\}$

Leave 2: $\{v_9, v_{10}, v_1\}$, $\{v_1, v_2, v_3\}$, $\{v_3, v_4, v_5\}$, $\{v_5, v_6, v_7\}$

Leave 3: $\{v_7, v_8, v_9\}$, $\{v_9, v_{10}, v_1\}$, $\{v_1, v_2, v_3\}$, $\{v_3, v_4, v_5\}$

Leave 4: $\{v_5, v_6, v_7\}$, $\{v_7, v_8, v_9\}$, $\{v_9, v_{10}, v_1\}$, $\{v_1, v_2, v_3\}$

Leave 5: $\{v_3, v_4, v_5\}$, $\{v_5, v_6, v_7\}$, $\{v_7, v_8, v_9\}$, $\{v_9, v_{10}, v_1\}$

The edges of the leaves can be partitioned into 4 copies of LC_5 when taken consecutively. Therefore, we have the desired LC_5-decomposition of $K_v^{(3)}$. □

Corollary 1 *Let λ and v be positive integers with $v \geq 10$ and $\lambda \not\equiv 0 \pmod 5$, and suppose t is the remainder after dividing λ by 5. When $v \equiv 3 \pmod 5$, the leave of a maximum LC_5-packing of ${}^\lambda K_v^{(3)}$ consists of t edges. When $v \equiv 4 \pmod 5$, the leave of a maximum LC_5-packing of ${}^\lambda K_v^{(3)}$ consists of $5 - t$ edges. In both cases, it is possible to configure the leave edges as a loose path.*

Proof First notice that ${}^\lambda K_v^{(3)}$ can be decomposed into copies of ${}^5 K_v^{(3)}$ and one copy of ${}^t K_v^{(3)}$. Furthermore, ${}^t K_v^{(3)}$ can be decomposed into t copies of $K_v^{(3)}$.

When $v \equiv 3 \pmod 5$ take a maximum LC_5-packing of $K_v^{(3)}$ with leave of a single edge, which exists by Theorem 2, on each of the t copies of $K_v^{(3)}$. These leave edges can be configured into any desired subgraph of $K_v^{(3)}$ with t edges, including

LP_t. Thus we have the desired LC_5-packing of ${}^{\lambda}K_v^{(3)}$, which we know is maximum due to the number of edges being less than 5.

When $v \equiv 4 \pmod 5$ take a maximum LC_5-packing of $K_v^{(3)}$ with leave of an LP_4, which exists by Theorem 2, on each of the t copies of $K_v^{(3)}$. We may choose distinct vertices $v_1, v_2, \ldots, v_{10}$ of $K_v^{(3)}$ so that the leaves of the t packings are Leaves $1, 2, \ldots, t$ as defined in the proof of Theorem 3. The sequence of the edges of these t leaves forms the following hypergraphs:

- when $t = 1$, we have a single LP_4 whose edges are the leave for our LC_5-packing of ${}^{t}K_v^{(3)}$;
- when $2 \leq t \leq 4$, we have $t - 1$ additional copies of LC_5 and the edges of an LP_{5-t} for the leave of our LC_5-packing of ${}^{t}K_v^{(3)}$.

Thus, we have the desired LC_5-packing of ${}^{\lambda}K_v^{(3)}$, which we know is maximum due to the number of edges in the leave being less than 5. □

Corollary 2 *Let λ and v be positive integers with $v \geq 10$ and $\lambda \not\equiv 0 \pmod 5$, and suppose t is the remainder after dividing λ by 5. When $v \equiv 3 \pmod 5$, a minimum LC_5-covering of ${}^{\lambda}K_v^{(3)}$ has a padding with $5 - t$ edges. When $v \equiv 4 \pmod 5$, a minimum LC_5-covering of ${}^{\lambda}K_v^{(3)}$ has a padding with t edges. In both cases, it is possible to configure the padding edges as a loose path.*

Proof When $v \equiv 3 \pmod 5$ take a maximum LC_5-packing of ${}^{\lambda}K_v^{(3)}$ with leave LP_t, which exists by Corollary 1. Since $t \leq 4$ and $v \geq 10$, this copy of LP_t can be completed to a copy of LC_5 (which contains 10 vertices) by attaching an appropriate copy of LP_{5-t}. This additional copy of LP_{5-t} is then the padding of our desired minimum LC_5-covering of ${}^{\lambda}K_v^{(3)}$, which we know is minimum due to the number of padding edges being less than 5.

When $v \equiv 4 \pmod 5$ take a maximum LC_5-packing of ${}^{\lambda}K_v^{(3)}$ with leave LP_{5-t}, which exists by Corollary 1. Similarly to the previous case, we can complete this LP_{5-t} into a copy of LC_5 by attaching a copy of LP_t, which becomes the padding of our desired minimum LC_5-covering of ${}^{\lambda}K_v^{(3)}$, which we know is minimum due to the number of padding edges being less than 5. □

Acknowledgments The authors would like to thank the anonymous referees, whose comments were very insightful and improved the readability of the paper.

This research is supported by grant number A1659815 from the Division of Mathematical Sciences at the National Science Foundation. This work was done while the second, third, fourth, and sixth authors were participants in *REU Site: Mathematics Research Experience for Pre-service and for In-service Teachers* at Illinois State University. The fifth and sixth authors were supported by a grant from Illinois Wesleyan University. The authors would like to thank Saad El-Zanati for his mentorship while conducting this research.

1 Introduction

Let $(G, \cdot)$ be a group. We continue the investigation of when the elements of a subset of G may be ordered in such a way that their partial sums are distinct. We set up a general framework for investigating semidirect product groups and look in detail at the dihedral groups of order twice a prime and non-abelian groups of order three times a prime.

Definition 1 Let S be a subset of the non-identity elements of a group G with $|S| = k$. If $x_1, x_2, \ldots, x_k$ is an ordering of the elements of S, define the jth *partial sum* of the ordering by $s_0 = id_G$ and $s_j = \prod_{i=1}^{j} x_i$ for $0 < j \leq k$. The ordering of the subset S is a *linear sequencing* if the partial sums $s_0, s_1, \ldots, s_k$ are all different. It is a *cyclic sequencing* if the partial sums are different with the exception that $s_0 = s_k = id_G$. If S has a linear or cyclic sequencing then it is *sequenceable*.

If every subset of the non-identity elements of a group is sequenceable then the group is *strongly sequenceable*. Combining several results and conjectures concering subsets of abelian groups dating back to 1961 [8], we have the conjecture that all abelian groups are strongly sequenceable. See [2, 7, 9, 18] for details of the history, motivation and current state of this conjecture.

The question for non-abelian groups is much more widely open. In [6, 16] small counterexamples are presented to show that the potential conjecture that all non-abelian groups are strongly sequenceable is false.

Keedwell's Conjecture [12] is that if S contains all the non-identity elements of a non-abelian group G, then a linear sequencing exists if and only if $|G| \geq 10$. This is known to hold for many groups, including dihedral groups [11, 13], soluble groups with a single involution [3], at least one group of every odd order at which a non-abelian group exists [17] and all groups of order at most 255 [15].

Let S be a subset of size k of $G \setminus \{id_G\}$ whose elements can be ordered so that their total sum is not the identity. Then S is known to have a linear sequencing in the following situations [16]:

- $k \leq 4$ and S does not have one of two specific structures,
- G is dihedral of order $2m$ for prime $m > 3$ and $k \leq 9$,
- G is dihedral of order $2m$ for even or prime $m > 4$ and $k = 2m - 2$.

In this paper we make two contributions to finding linear sequencings in non-abelian groups. First, we improve the second item above, showing that it holds up to $k = 12$ and also for m whose prime factors are each larger than $k!$. (The result that allows for composite m with large prime factors in fact applies more widely than dihedral groups; see Sect. 3.) The second contribution concerns non-abelian groups of order three times a prime p; there is at most one such group for any p and we denote it G_{3p}. In this case we show that subsets S of size $5 < k \leq 10$ are sequenceable and, further, that we can always find a linear sequencing for these subsets when the sum of all the elements is not unavoidably the identity.

In [4] a generalization of sequenceability is introduced.

Definition 2 Let S be a subset of the non-identity elements of a group G with $|S| = k$ and let $x_1, x_2, \ldots, x_k$ be an ordering of the elements of S with partial sums $s_0, s_1, \ldots, s_k$. If $s_i \neq s_j$ for any distinct pair of indices i, j with $|i - j| \leq t$ then the ordering is a *t-weak sequencing* of S and S is *t-weakly sequenceable.*

Observe that a k-weak sequencing is exactly a linear sequencing and that a $(k - 1)$-weak sequencing is either a linear or cyclic sequencing.

In [4] it is shown that a subset of non-identity elements of the cyclic group $\mathbb{Z}_p$ of prime order p is t-weak sequenceable when $t \leq 6$ and, with the additional condition that there are no pairs of the form $\{x, -x\}$ in the subset, when $t = 7$. We extend the methodology to apply to groups G that are not cyclic of prime order, showing that if $|G| = pe$ then all sufficiently large subsets of $G \setminus \{id_G\}$ are t-weakly sequenceable when $p > 3$ is prime, $e \leq 3$ and $t \leq 6$.

In the next section we describe the polynomial method and how it can be implemented for groups that are semidirect products. In Sects. 2.1 and 3 we apply the method to obtain the results on sequenceability and in Sect. 4 we apply it to obtain the results on weak sequenceability.

2 Polynomial Method in Semidirect Products

In this section, we apply a method that relies on the Non-Vanishing Corollary of the Combinatorial Nullstellensatz, see [1, 14]. Given a prime p (in the following p will be always assumed to be a prime), this corollary allows us to obtain a nonzero point to suitable polynomials on $\mathbb{Z}_p$ derived starting from the ones defined in [10] and in [16].

Theorem 1 (Non-Vanishing Corollary) *Let $\mathbb{F}$ be a finite field, and let $f(x_1, x_2, \ldots, x_k)$ be a polynomial in $\mathbb{F}[x_1, x_2, \ldots, x_k]$. Suppose the degree $deg(f)$ of f is $\sum_{i=1}^{k} \gamma_i$, where each γ_i is a nonnegative integer, and suppose the coefficient of $\prod_{i=1}^{k} x_i^{\gamma_i}$ in f is nonzero. If $C_1, C_2, \ldots, C_k$ are subsets of $\mathbb{F}$ with $|C_i| > \gamma_i$, then there are $c_1 \in C_1, \ldots, c_k \in C_k$ such that $f(c_1, c_2, \ldots, c_k) \neq 0$.*

In the notation of the Non-Vanishing Corollary, we call the monomial $x_1^{|C_1|-1} \cdots x_k^{|C_k|-1}$ the *bounding monomial.* The corollary can be rephrased as requiring the polynomial to include a monomial of maximum degree that divides the bounding monomial (where by "include" we mean that it has a nonzero coefficient).

To use the Non-Vanishing Corollary we require a polynomial for which the nonzeros correspond to successful solutions to the case of the problem under consideration. Here, we will consider subsets S of a group $(G, \cdot_G)$ that is the semidirect product of $(\mathbb{Z}_p, +)$ and another group $(H, \cdot_H)$. For such groups we will use the notation $G = \mathbb{Z}_p \rtimes_\varphi H$ for some group homomorphism $\varphi : H \to Aut(\mathbb{Z}_p)$.

Remark 1 These kind of semidirect products define a wide class of groups. Our hypotheses are, indeed, equivalent to consider groups $(G, \cdot)$ that admit an isomorphic copy of $(\mathbb{Z}_p, +)$ as a normal subgroup N and whose cosets are of the form $N \cdot id, N \cdot c_1, \ldots, N \cdot c_{e-1}$ where $e = |G/N|$ and $\{c_0 = id, \ldots, c_{e-1}\}$ is a subgroup of G.

We now revisit the polynomial method introduced in [16] for dihedral groups and in [7] for groups of type $\mathbb{Z}_p \times H$. To avoid confusion, from now on, the symbol $+$ will be used only between elements of $\mathbb{Z}_p$. Let $\pi_2 : \mathbb{Z}_p \rtimes_\varphi H \to H$ be the projection map that picks out the second coordinate of an element and for a subset $S \subseteq (\mathbb{Z}_p \rtimes_\varphi H) \setminus \{id_{\mathbb{Z}_p \rtimes_\varphi H}\}$ let $\pi_2(S)$ be the multiset $\{\pi_2(s) : s \in S\}$. Also, the generic element of $\mathbb{Z}_p \rtimes_\varphi H$ will be indicated by $x \cdot_G a$ (or $y \cdot b$) implying that $x \in \mathbb{Z}_p$ and $a \in H$. We define the *type* of S to be the sequence $\boldsymbol{\lambda} = (\lambda_0, \ldots, \lambda_{e-1})$, where λ_i is the number of times that c_i appears in $\pi_2(S)$.

Let T be a multiset of elements from $H \setminus \{id_H\}$ with size k. Let $\boldsymbol{a} = (a_1, \ldots, a_k)$ be an arrangement of the elements of T with partial sums $\boldsymbol{b} = (b_0, b_1, \ldots, b_k)$. If, for each i, the element c_i appears at most r times in $\mathbf{b}$, then $\mathbf{a}$ is a *quotient sequencing* of T with respect to r. In our setting, given some $S \subseteq (\mathbb{Z}_p \rtimes_\varphi H) \setminus \{id_{\mathbb{Z}_p \rtimes_\varphi H}\}$ for which we wish to find a sequencing, we shall be interested in quotient sequencings of $\pi_2(S)$ with respect to p.

Given $S \subseteq (\mathbb{Z}_p \rtimes_\varphi H) \setminus \{id_{\mathbb{Z}_p \rtimes_\varphi H}\}$ we first construct a quotient sequencing of $\pi_2(S)$ with respect to p. We then use the polynomial method to show that there is a sequencing for S that projects elementwise onto that quotient sequencing. Given such a quotient sequencing $\boldsymbol{a} = (a_1, \ldots a_k)$ with partial sums $\boldsymbol{b} = (b_0, b_1, \ldots, b_k)$, let

$$\boldsymbol{x_a} = (x_1 \cdot_G a_1, x_2 \cdot_G a_2, \ldots, x_k \cdot_G a_k)$$

be a putative arrangement of the elements of S with partial sums

$$\boldsymbol{y_a} = (s_0, \ldots, s_k) = (y_0 \cdot_G b_0, y_1 \cdot_G b_1, \ldots, y_k \cdot_G b_k)\,.$$

We will need the following lemma:

Lemma 1 *Let $(G, \cdot_G)$ be a group of type $\mathbb{Z}_p \rtimes_\varphi H$. Then, considering $x_1 \cdot_G a_1$ and $x_2 \cdot_G a_2$ in G, there exists $\varphi_{a_1} \in \mathbb{Z}_p$ such that:*

$$(x_1 \cdot_G a_1) \cdot_G (x_2 \cdot_G a_2) = (x_1 + (\varphi_{a_1} x_2)) \cdot_G (a_1 \cdot_H a_2)$$

where $\varphi_{a_1} x_2$ represent the product between the elements of $\mathbb{Z}_p$, φ_{a_1} and x_2.

Proof Due to the definition, we have that

$$(x_1 \cdot_G a_1) \cdot_G (x_2 \cdot_G a_2) = (x_1 + \varphi(a_1)(x_2)) \cdot_G (a_1 \cdot_H a_2).$$

Here the map

$$x \to \varphi(a_1)(x)$$

is an automorphism of $(\mathbb{Z}_p, +)$. It is well known that the automorphism of $(\mathbb{Z}_p, +)$ are the multiplications and hence there exists $\varphi_{a_1} \in \mathbb{Z}_p$ such that

$$\varphi(a_1)(x) = \varphi_{a_1} x.$$

Hence we have that

$$(x_1 + \varphi(a_1)(x_2)) \cdot_G (a_1 \cdot_H a_2) = (x_1 + (\varphi_{a_1} x_2)) \cdot_G (a_1 \cdot_H a_2)$$

which implies the thesis. □

From now on, when it is clear from the context, the index G or H in the products will be omitted.

Here we note that, if $j > i$, $s_i = s_j$ only when

$$(x_{i+1} \cdot a_{i+1}) \cdot (x_{i+2} \cdot a_{i+2}) \cdot \ldots \cdot (x_j \cdot a_j) = id.$$

Applying Lemma 1, this equality can be written as

$$(x_{i+1} + \varphi_{a_{i+1}} x_{i+2}) \cdot (a_{i+1} \cdot a_{i+2}) \cdot \ldots \cdot (x_j \cdot a_j) = id$$

and, reiterating it, as

$$(x_{i+1} + \varphi_{a_{i+1}} x_{i+2} + \varphi_{a_{i+1} \cdot a_{i+2}} x_{i+3} + \cdots + \varphi_{a_{i+1} \ldots a_{j-1}} x_j) \cdot (a_{i+1} \cdot a_{i+2} \cdot \ldots \cdot a_j) = id.$$

This implies that

$$a_{i+1} \cdot a_{i+2} \cdot \ldots \cdot a_j = id$$

which is $b_i = b_j$, and

$$x_{i+1} + \varphi_{a_{i+1}} x_{i+2} + \varphi_{a_{i+1} \cdot a_{i+2}} x_{i+3} + \cdots + \varphi_{a_{i+1} \ldots a_{j-1}} x_j = 0. \tag{1}$$

Hence, in order to impose that $s_i \neq s_j$ for any $i \neq j$ is enough to consider Eq. (1) for the pairs for which $b_i = b_j$ (see also [7]). Therefore for a fixed quotient sequencing **a**, we can consider the following polynomial in the variables $x_1, x_2, \ldots, x_k$

$$p_{\mathbf{a}} = \prod_{\substack{1 \leq i < j \leq k \\ a_i = a_j}} (x_j - x_i) \times \prod_{\substack{0 \leq i < j \leq k \\ b_i = b_j \\ j \neq i+1}} (x_{i+1} + \varphi_{a_{i+1}} x_{i+2} + \varphi_{a_{i+1} \cdot a_{i+2}} x_{i+3} + \cdots + \varphi_{a_{i+1} \ldots a_{j-1}} x_j) \tag{2}$$

Table 1 Monomials and their coefficients in the case $|S| = 12$ and $G = D_{2p}$

$\boldsymbol{\lambda}$	**a**	Deg	Monomial/s	Coefficient/s
(11, 1)	(0, 0, 0, 0, 0, 1, 0, 0, 0, 0, 0, 0)	71	$x_2^3x_3^5x_4^7x_5^9x_7^9x_8^9x_9^9x_{10}^9x_{11}^{10}x_{12}^{10}$ $x_2^3x_3^5x_4^8x_5^8x_7^9x_8^9x_9^9x_{10}^9x_{11}^{10}x_{12}^{10}$	$2 \cdot 3 \cdot 7^2 \cdot 131$ $2 \cdot 5 \cdot 7 \cdot 593$
(10, 2)	(0, 0, 0, 0, 0, 1, 0, 0, 0, 0, 0, 1)	64	$x_2^4x_3^5x_4^9x_5^8x_6x_7^8x_8^9x_9^9x_{10}^9x_{11}^9x_{12}$ $x_2^4x_3^5x_4^8x_5^9x_6x_7^8x_8^9x_9^9x_{10}^9x_{11}^9x_{12}$	$-2^7 \cdot 3 \cdot 29 \cdot 41$ $-2^5 \cdot 3 \cdot 101 \cdot 227$
(9, 3)	(0, 0, 0, 0, 1, 1, 1, 0, 0, 0, 0, 0)	57	$x_2^2x_3^8x_4^8x_5^2x_6^2x_7^2x_8^8x_9^8x_{10}^8x_{11}^8x_{12}^8$ $x_2^3x_3^7x_4^8x_5^2x_6^2x_7^2x_8^8x_9^8x_{10}^8x_{11}^8x_{12}^8$	$-2^3 \cdot 3^3 \cdot 5 \cdot 31$ $-2^2 \cdot 3^2 \cdot 61 \cdot 73$
(8, 4)	(0, 0, 0, 0, 1, 1, 1, 0, 0, 0, 0, 1)	54	$x_2^6x_3^7x_4^7x_5^3x_6^3x_7^3x_8^7x_9^7x_{10}^7x_{11}^7x_{12}^3$ $x_2^7x_3^7x_4^7x_5^3x_6^3x_7^3x_8^6x_9^7x_{10}^7x_{11}^7x_{12}^3$	$2^6 \cdot 7 \cdot 7669$ $-2^6 \cdot 11 \cdot 28591$
(7, 5)	(0, 0, 0, 1, 1, 1, 1, 1, 0, 0, 0, 0)	51	$x_2^6x_3^6x_4^4x_5^4x_6^4x_7^4x_8^4x_9^6x_{10}^6x_{11}^6x_{12}^6$ $x_1^6x_3^6x_4^4x_5^4x_6^4x_7^4x_8^4x_9^6x_{10}^6x_{11}^6x_{12}^6$	$-2^3 \cdot 3^2 \cdot 5 \cdot 7 \cdot 11 \cdot 401$ $-2^2 \cdot 5^2 \cdot 179 \cdot 389$
(6, 6)	(0, 0, 0, 1, 1, 1, 1, 1, 0, 0, 0, 1)	52	$x_1^2x_2^4x_3^5x_4^5x_5^5x_6^5x_7^5x_8^5x_9^5x_{10}^5x_{11}^5x_{12}^5$ $x_1^3x_2^3x_3^5x_4^5x_5^5x_6^5x_7^5x_8^5x_9^5x_{10}^5x_{11}^5x_{12}^5$	$2 \cdot 3^3 \cdot 181 \cdot 481373$ $-2 \cdot 3^2 \cdot 13 \cdot 22399121$
(4, 8)	(0, 0, 1, 1, 1, 1, 1, 1, 1, 0, 0, 1)	58	$x_2^3x_3^4x_4^5x_5^7x_6^7x_7^7x_8^7x_9^7x_{10}^3x_{11}^3x_{12}^7$ $x_2^3x_3^4x_4^6x_5^6x_6^7x_7^7x_8^7x_9^7x_{10}^3x_{11}^3x_{12}^7$	$2^4 \cdot 3 \cdot 79 \cdot 179 \cdot 18979$ $-2^4 \cdot 3^2 \cdot 5 \cdot 109 \cdot 119839$
(3, 9)	(0, 1, 1, 1, 1, 1, 1, 1, 1, 1, 0, 0)	63	$x_1^2x_2^4x_3^6x_4^8x_5^8x_7^8x_8^8x_9^8x_{10}^8x_{11}^2x_{12}^2$ $x_1^2x_2^5x_3^5x_4^8x_5^8x_7^8x_8^8x_9^8x_{10}^8x_{11}^2x_{12}^2$	$3 \cdot 62701 \cdot 131111$ $-2 \cdot 3 \cdot 5 \cdot 13 \cdot 17 \cdot 3699679$
(2, 10)	(0, 1, 1, 1, 1, 1, 1, 1, 1, 1, 0, 1)	72	$x_1x_2^4x_3^4x_4^8x_5^9x_7^9x_8^9x_9^9x_{10}^9x_{11}x_{12}^9$ $x_1x_2^4x_3^5x_4^7x_5^9x_7^9x_8^9x_9^9x_{10}^9x_{11}x_{12}^9$	$2 \cdot 5 \cdot 71 \cdot 1427 \cdot 1033463$ $-2 \cdot 3^2 \cdot 7 \cdot 23 \cdot 215947813$
(1, 11)	(1, 1, 1, 1, 1, 1, 1, 1, 1, 1, 1, 0)	80	$x_2^5x_3^5x_4^7x_5^7x_6^8x_7^8x_8^{10}x_9^{10}x_{10}^{10}x_{11}^{10}$ $x_2^4x_3^6x_4^7x_5^7x_6^8x_7^8x_8^{10}x_9^{10}x_{10}^{10}x_{11}^{10}$	$2^3 \cdot 7 \cdot 101 \cdot 81083329$ $-2^5 \cdot 5 \cdot 31 \cdot 3557 \cdot 21647$

Table 2 Monomials and their coefficients in the case $|S| = 10$ and $G = G_{3p}$

λ	a	Deg	Monomial/s	Coefficient/s
(9, 1, 0)	(0, 0, 0, 0, 0, 1, 0, 0, 0, 0)	52	$x_2^2x_3^4x_4^7x_5^8x_7^7x_8^8x_9^8x_{10}^8$	-2^2
(8, 2, 0)	(0, 0, 0, 1, 0, 0, 0, 1, 0, 0)	36	$x_2^2x_3^4x_5^3x_6^6x_7^7x_8x_9^6x_{10}^7$	-2
(8, 1, 1)	(0, 0, 0, 0, 1, 0, 0, 0, 0, 2)	45	$x_1x_2^4x_3^6x_4^7x_6^6x_7^7x_8^7x_9^7$	$-1791r+573$
			$x_1^2x_2^3x_3^6x_4^7x_6^6x_7^7x_8^7x_9^7$	$-235r-746$
(7, 3, 0)	(0, 0, 1, 0, 0, 0, 1, 0, 0, 1)	32	$x_1^2x_2^3x_4^2x_5^5x_6^6x_7x_8^5x_9^6x_{10}^2$	$14r+73$
			$x_1^2x_2^2x_4^3x_5^5x_6^6x_7x_8^5x_9^6x_{10}^2$	$12r+21$
(7, 2, 1)	(0, 0, 0, 1, 0, 0, 1, 0, 0, 2)	30	$x_1x_2^2x_3^4x_4x_5^5x_6^5x_7x_8^5x_9^6$	1
(6, 4, 0)	(0, 0, 1, 0, 0, 1, 0, 0, 1, 1)	30	$x_2^2x_3^2x_4^4x_5^5x_6^2x_7^4x_8^5x_9^3x_{10}^3$	$-1168r+72$
			$x_1x_2^2x_3^2x_4^3x_5^5x_6^2x_7^4x_8^5x_9^3x_{10}^3$	$-436r-1078$
(6, 3, 1)	(0, 0, 0, 1, 0, 0, 1, 0, 2, 1)	27	$x_2^2x_3^5x_4x_5^5x_6^5x_7^2x_8^5x_{10}^2$	$r+1$
(6, 2, 2)	(0, 0, 1, 0, 0, 1, 0, 0, 2, 2)	26	$x_1^3x_2^3x_3x_4^3x_5^4x_6x_7^4x_8^5x_9x_{10}$	$10r-64$
			$x_1^2x_2^5x_4^3x_5^4x_6x_7^4x_8^5x_9x_{10}$	$155r+583$
(5, 5, 0)	(0, 0, 1, 0, 0, 1, 0, 1, 1, 1)	30	$x_2^2x_3x_4^3x_5^4x_6^4x_7^4x_8^4x_9^4x_{10}^4$	$1965r+2715$
			$x_1x_2^2x_3^2x_4^2x_5^3x_6^4x_7^4x_8^4x_9^4x_{10}^4$	$385r-190$
(5, 4, 1)	(0, 0, 1, 0, 0, 1, 0, 2, 1, 1)	26	$x_1x_2^4x_3x_4^3x_5^4x_6^3x_7^4x_9^3x_{10}^3$	$30r+156$
			$x_1^2x_2^3x_3x_4^3x_5^4x_6^3x_7^4x_9^3x_{10}^3$	$42r+72$
(5, 3, 2)	(0, 0, 0, 1, 0, 1, 0, 2, 1, 2)	24	$x_1^4x_2^4x_3^3x_4^2x_5^4x_6^2x_7^2x_8x_9^2$	2
(4, 6, 0)	(0, 1, 0, 0, 1, 0, 1, 1, 1, 1)	32	$x_1^3x_2^5x_3^3x_4^3x_5^5x_6^3x_7^4x_8^4x_9x_{10}$	$-11377r+11374$
			$x_1^2x_2^5x_3^3x_4^3x_5^5x_6^3x_7^5x_8^4x_9x_{10}$	$-21587r-27358$
(4, 5, 1)	(0, 0, 1, 0, 1, 0, 1, 1, 1, 2)	27	$x_2^2x_3^3x_4^3x_5^4x_6^3x_7^4x_8^4x_9^4$	$-1286r+68$
			$x_1^2x_2^2x_3^3x_4^3x_5^3x_6^3x_7^3x_8^4x_9^4$	$-333r-119$
(4, 4, 2)	(0, 0, 0, 1, 0, 1, 2, 1, 2, 1)	24	$x_1^2x_2^2x_3^3x_4^3x_5^3x_6^3x_7x_8^3x_9x_{10}^3$	$-2r$
(4, 3, 3)	(0, 0, 1, 0, 1, 0, 2, 1, 2, 2)	23	$x_1^2x_2^3x_3^2x_4^3x_5^2x_6^3x_7^2x_8^2x_9^2x_{10}^2$	$-1452r+1686$
			$x_1^3x_2^3x_3^2x_4^3x_5^2x_6^2x_7^2x_8^2x_9^2x_{10}^2$	$-759r-348$
(3, 7, 0)	(0, 1, 0, 1, 0, 1, 1, 1, 1, 1)	36	$x_2x_3^2x_4x_5^2x_6^6x_7^6x_8^6x_9^6x_{10}^6$	$-19616r+331264$
			$x_2x_3^2x_4^3x_5^2x_6^4x_7^6x_8^6x_9^6x_{10}^6$	$802469r-381850$
(3, 6, 1)	(0, 0, 1, 0, 1, 1, 1, 1, 2, 1)	30	$x_1^2x_2x_3^5x_4^2x_5^5x_6^5x_7^5x_8^4x_{10}$	$3396r-230$
			$x_1^2x_2x_3^3x_4^2x_5^5x_6^5x_7^5x_8^4x_{10}^3$	$133r+344$
(3, 5, 2)	(0, 0, 1, 0, 1, 2, 1, 2, 1, 1)	26	$x_1^2x_2^2x_3^2x_4^2x_5^4x_6x_7^4x_8x_9^4x_{10}^4$	$-3r+24$
			$x_1^2x_2^2x_3^3x_4^2x_5^3x_6x_7^4x_8x_9^4x_{10}^4$	$21r+21$
(3, 4, 3)	(0, 0, 1, 0, 1, 1, 2, 2, 1, 2)	24	$x_1^2x_2^2x_3^3x_4^2x_5^3x_6^3x_7^2x_8^2x_9^3x_{10}^2$	$1358r+1812$
	(0, 0, 1, 1, 0, 1, 2, 2, 1, 2)	24	$x_1^2x_2^2x_3^3x_4^3x_5^2x_6^3x_7^2x_8^2x_9^3x_{10}^2$	$-1742r-1035$
(2, 8, 0)	(0, 1, 0, 1, 1, 1, 1, 1, 1, 1)	42	$x_3x_4^2x_5^4x_6^7x_7^7x_8^7x_9^7x_{10}^7$	$65374r+67999$
			$x_3x_4^2x_5^5x_6^6x_7^7x_8^7x_9^7x_{10}^7$	$-160333r-7057$

(continued)

Table 2 (continued)

λ	**a**	Deg	Monomial/s	Coefficient/s
(2, 7, 1)	(0, 1, 0, 1, 1, 1, 1, 2, 1, 1)	35	$x_1 x_3 x_4^3 x_5^6 x_6^6 x_7^6 x_9^6 x_{10}^6$	$17861r - 27263$
			$x_1 x_3 x_4^4 x_5^5 x_6^6 x_7^6 x_9^6 x_{10}^6$	$12821r + 26143$
(2, 6, 2)	(2, 0, 1, 1, 1, 1, 1, 2, 1, 0)	30	$x_3^3 x_4^5 x_5^5 x_6^5 x_7^5 x_8 x_9^5 x_{10}$	$-292r + 246$
			$x_3^4 x_4^4 x_5^5 x_6^5 x_7^5 x_8 x_9^5 x_{10}$	$85r + 126$
(2, 5, 3)	(2, 1, 1, 1, 0, 2, 2, 1, 0, 1)	27	$x_1^2 x_2^4 x_3^4 x_4^4 x_6^2 x_7^2 x_8^4 x_9 x_{10}^4$	$17651r - 23546$
			$x_1^2 x_2^3 x_3^4 x_4^4 x_5 x_6^2 x_7^2 x_8^4 x_9 x_{10}^4$	$-10860r - 152465$
(2, 4, 4)	(0, 1, 0, 1, 1, 2, 2, 1, 2, 2)	26	$x_1 x_2^3 x_3 x_4^3 x_5^3 x_6^3 x_7^3 x_8^3 x_9^3 x_{10}^3$	$136392r + 247068$
	(0, 1, 1, 0, 1, 2, 2, 1, 2, 2)	26	$x_1 x_2^3 x_3^3 x_4 x_5^3 x_6^3 x_7^3 x_8^3 x_9^3 x_{10}^3$	$-175716r - 167652$
(1, 9, 0)	(1, 0, 1, 1, 1, 1, 1, 1, 1, 1)	50	$x_3^2 x_4^4 x_5^5 x_6^7 x_7^8 x_8^8 x_9^8 x_{10}^8$	$992906r + 1427608$
			$x_3^3 x_4^3 x_5^5 x_6^7 x_7^8 x_8^8 x_9^8 x_{10}^8$	$-1815167r - 733348$
(1, 8, 1)	(0, 1, 1, 1, 1, 1, 1, 1, 1, 2)	42	$x_2 x_3^2 x_4^4 x_5^7 x_6^7 x_7^7 x_8^7 x_9^7$	$63854r - 6907$
			$x_2 x_3^2 x_4^5 x_5^6 x_6^7 x_7^7 x_8^7 x_9^7$	$38518r + 68818$
(1, 7, 2)	(0, 1, 1, 1, 1, 1, 1, 2, 2, 1)	36	$x_3^4 x_4^6 x_5^6 x_6^6 x_7^6 x_8 x_9 x_{10}^6$	$-16124r - 27940$
			$x_3^5 x_4^5 x_5^6 x_6^6 x_7^6 x_8 x_9 x_{10}^6$	$3718r + 13449$
(1, 6, 3)	(1, 1, 0, 1, 1, 1, 2, 1, 2, 2)	32	$x_1 x_2^5 x_4^5 x_5^5 x_6^5 x_7^2 x_8^5 x_9^2 x_{10}^2$	$31065r + 30039$
			$x_1^2 x_2^4 x_4^5 x_5^5 x_6^5 x_7^2 x_8^5 x_9^2 x_{10}^2$	$1764r + 9147$
(1, 5, 4)	(0, 1, 1, 1, 1, 1, 2, 2, 2, 2)	30	$x_2^4 x_3^4 x_4^4 x_5^4 x_6^4 x_7^3 x_8^3 x_9^3 x_{10}$	$211692r + 376028$
			$x_2^2 x_3^4 x_4^4 x_5^4 x_6^4 x_7^3 x_8^3 x_9^3 x_{10}^3$	$78700r - 731650$
(0, 10, 0)	(1, 1, 1, 1, 1, 1, 1, 1, 1, 1)	60	$x_1^2 x_2^2 x_3^4 x_4^7 x_5^9 x_7^9 x_8^9 x_9^9 x_{10}^9$	$-250444r - 433547$
			$x_1^2 x_2^2 x_3^5 x_4^6 x_5^9 x_7^9 x_8^9 x_9^9 x_{10}^9$	$1040607r + 497898$
(0, 9, 1)	(1, 1, 1, 1, 1, 2, 1, 1, 1, 1)	51	$x_2^2 x_3^3 x_4^6 x_5^8 x_7^8 x_8^8 x_9^8 x_{10}^8$	$21957r - 14546$
			$x_2^2 x_3^4 x_4^5 x_5^8 x_7^8 x_8^8 x_9^8 x_{10}^8$	$37855r + 47642$
(0, 8, 2)	(1, 1, 1, 1, 1, 1, 1, 1, 2, 2)	44	$x_2^3 x_3^4 x_4^7 x_5^7 x_6^7 x_7^7 x_8^7 x_9 x_{10}$	$1698588r + 406932$
			$x_2^3 x_3^5 x_4^6 x_5^7 x_6^7 x_7^7 x_8^7 x_9 x_{10}$	$201528r + 1213386$
(0, 7, 3)	(2, 1, 1, 1, 1, 1, 1, 1, 2, 2)	39	$x_1 x_2^2 x_3^2 x_4^6 x_5^6 x_6^6 x_7^6 x_8^6 x_9^2 x_{10}^2$	$824103r + 258808$
			$x_1 x_2^2 x_3^3 x_4^5 x_5^6 x_6^6 x_7^6 x_8^6 x_9^2 x_{10}^2$	$89629r + 972050$
(0, 6, 4)	(2, 1, 2, 1, 1, 1, 1, 1, 2, 2)	38	$x_1 x_2^3 x_3^3 x_4^5 x_5^5 x_6^5 x_7^5 x_8^5 x_9^3 x_{10}^3$	$1390432r + 1536910$
			$x_2^4 x_3^3 x_4^5 x_5^5 x_6^5 x_7^5 x_8^5 x_9^3 x_{10}^3$	$-792953r - 108625$
(0, 5, 5)	(1, 1, 1, 1, 1, 2, 2, 2, 2, 2)	35	$x_1 x_2^2 x_3^4 x_4^4 x_5^4 x_6^4 x_7^4 x_8^4 x_9^4 x_{10}^4$	$-710283r + 3022047$
			$x_1^2 x_2^2 x_3^3 x_4^4 x_5^4 x_6^4 x_7^4 x_8^4 x_9^4 x_{10}^4$	$2272296r + 657732$

$$b^2 - ab + a^2 \equiv 0 \pmod{p}. \tag{3}$$

On the other hand, if $a \equiv 0 \pmod{p}$ the coefficient is zero only when also b is zero modulo p and hence Eq. (3) is satisfied also in this case. Since Eq. (3) is satisfied for finitely many p, one can check that, for any prime p, at least one of the coefficients presented in Table 2 below is nonzero. □

Remark 2 Since in Eq. (2) we have imposed also $s_0 \neq s_k$, we have that in case $G = D_{2p}$ (resp. $G = G_{3p}$), S is not of type $(k, 0)$ (resp. $(k, 0, 0)$) and k satisfies the hypotheses of the previous theorem, then S admits a linear sequencing.

This implies that, under the hypothesis of Theorem 2, S admits a linear sequencing unless s_k is unavoidably the identity.

The following Tables 1 and 2 contain the required monomials and their coefficients for the cases $(G, k) = (D_{2p}, 12)$ and $(G, k) = (G_{3p}, 10)$.

3 To Infinity and Back

In this section we apply a method of [5] and [7], outlined also in [20], to generalize the sequenceability results here obtained to a more general class of semidirect products.

First, given a subset S of a group G, we define the set

$$\Upsilon(S) = S \cup \Delta(S),$$

where $\Delta(S) = \{x_1 - x_2 \mid x_1, x_2 \in S, x_1 \neq x_2\}$.

Then we can adapt [5, Lemma 2.1] as

Lemma 2 *Let G_1 and G_2 be groups. Given a subset S of $G_1 \setminus \{id_{G_1}\}$ of size k, suppose there exists a homomorphism $\phi : G_1 \to G_2$ such that $\ker(\phi) \cap \Upsilon(S) = \varnothing$. Then $\phi(S)$ is a subset of $G_2 \setminus \{id_{G_2}\}$ of size k and the subset S is sequenceable whenever $\phi(S)$ is sequenceable.*

In the following, we consider groups of the form $(G \rtimes_\varphi H, \cdot)$ where $(G, +)$ is abelian, $H = \{h_0 = id_H, h_1, \ldots, h_{e-1}\}$ is finite, $(H, \cdot_H)$[1] is not necessarily abelian and where φ is a group homomorphism $\varphi : H \to Aut(G)$. We will also assume that the homomorphism φ satisfies the following property:

$(*)$ given $h \in H$, $\varphi(h)$ is either the identity or $\varphi(h)(x) = -x$ for any $x \in G$.

Note that the latter is an automorphism of G because G is abelian.

[1] Also in this section, when it is clear from the context, the index H will be omitted.

Remark 5 Because of Remark 2, with essentially the same proof of Theorem 5, we can prove that if S is a subset of $Dih(G)$ not of type $(k, 0)$, $k \leq 12$ and $\vartheta(G) > N(k)$, then S admits a linear sequencing.

This implies that, under the hypothesis of Theorem 5, S admits a linear sequencing unless s_k is unavoidably the identity.

3.1 An Explicit Upper Bound

Now, in case $G = \mathbb{Z}_m$, and adapting again the procedure of [7], we provide an explicit upper bound on the numbers $N(\boldsymbol{\lambda})$ and $N(k)$. First, we need to recall some elementary facts about the solutions of a linear system over fields and commutative rings.

Theorem 6 (Corollary of Cramer's Theorem) *Let*

$$\begin{cases} m_{1,1}x_1 + \cdots + m_{1,k}x_k = b_1 \\ \dots \\ m_{l,1}x_1 + \cdots + m_{l,k}x_k = b_l \end{cases} \tag{4}$$

be a linear system over a commutative ring R whose associated matrix $M = (m_{i,j})$ is an $l \times k$ matrix over R. Let $(x_1, \dots, x_k)$ be a solution of the system and let M' be a $k' \times k'$ square, nonsingular (i.e. $\det(M')$ *is invertible in R), submatrix of M. We assume, without loss of generality, that M' is obtained by considering the first k' rows and the first k' columns of M. Then $x_1, \dots, x_k$ satisfy the following relation*

$$\begin{bmatrix} x_1 \\ x_2 \\ \vdots \\ x_{k'} \end{bmatrix} = \begin{bmatrix} m_{1,1} & m_{1,2} & m_{1,3} & \dots & m_{1,k'} \\ m_{2,1} & m_{2,2} & m_{2,3} & \dots & m_{2,k'} \\ \vdots & \vdots & \vdots & & \vdots \\ m_{k',1} & m_{k',2} & m_{k',3} & \dots & m_{k',k'} \end{bmatrix}^{-1}$$

$$\times \begin{bmatrix} b_1 - (m_{1,k'+1}x_{k'+1} + \cdots + m_{1,k}x_k) \\ b_2 - (m_{2,k'+1}x_{k'+1} + \cdots + +m_{2,k}x_k) \\ \vdots \\ b_{k'} - (m_{k',k'+1}x_{k'+1} + \cdots + m_{k',k}x_k) \end{bmatrix}.$$

Equivalently, set

$$M'_j := \begin{bmatrix} m_{1,1} & m_{1,2} & \dots & m_{1,j-1} & b_1 - (m_{1,k'+1}x_{k'+1} + \cdots + m_{1,k}x_k) & \dots & m_{1,k'} \\ m_{2,1} & m_{2,2} & \dots & m_{2,j-1} & b_2 - (m_{2,k'+1}x_{k'+1} + \cdots + +m_{2,k}x_k) & \dots & m_{2,k'} \\ \vdots & \vdots & \vdots & \vdots & \vdots & & \vdots \\ m_{k',1} & m_{k',2} & \dots & m_{k',j-1} & b_{k'} - (m_{k',k'+1}x_{k'+1} + \cdots + m_{k',k}x_k) & \dots & m_{k',k'} \end{bmatrix}$$

we have that, for any $j \in [1, k']$

$$x_j = \frac{\det(M'_j)}{\det(M')}. \tag{5}$$

Now assume that R is a field and that the system (4) admits at least one solution. Then, given a maximal square nonsingular submatrix M' of M, and fixed $x_{k'+1}, \ldots, x_k$ in R, we have that $x_1, x_2, \ldots, x_k$ are a solution of the system (4) if and only if $x_1, \ldots, x_{k'}$ satisfy the relation (5).

Remark 6 In case the system (4) of Theorem 6 is homogeneous (i.e. $b_1 = b_2 = \cdots = b_l = 0$), due to the Laplace expansion of the determinant, $\det(M'_j)$ can be written as a linear combination of $x_{k'+1}, \ldots, x_k$. More precisely, denoting by $M'_{i,j}$ the matrix obtained by deleting the i-th row and the j-th column of M', we have that

$$\det(M'_j) = \sum_{i=1}^{k'} (-1)^{j+i+1} \det(M'_{i,j})(m_{i,k'+1}x_{k'+1} + \cdots + m_{i,k}x_k). \tag{6}$$

Now we are ready to provide, in case of groups of type $\mathbb{Z}_m \rtimes_\varphi H$ (where φ satisfies property $(*)$), a quantitative version of Theorem 4. In the following proof, developing the notations of the previous sections, given a set S of size k and an ordering $\boldsymbol{x} = (x_1, \ldots, x_k)$ of S, we denote by $s_i(\boldsymbol{x})$ the partial sum $x_1 \cdot x_2 \ldots x_{i-1} \cdot x_i$ where $s_0(\boldsymbol{x})$ is always zero.

Theorem 7 *Let H be a finite group, let φ satisfy property $(*)$, and suppose that, for infinitely many odd primes p, any subset of a given type $\boldsymbol{\lambda} = (\lambda_0, \ldots, \lambda_{e-1})$ of $\mathbb{Z}_p \rtimes_\varphi H \setminus \{id_{\mathbb{Z}_p \rtimes_\varphi H}\}$ is sequenceable. Then any subset of type $\boldsymbol{\lambda}$ of $\mathbb{Z}_m \rtimes_\varphi H \setminus \{id_{\mathbb{Z}_m \rtimes_\varphi H}\}$ is sequenceable provided that the prime factors of m are all greater than $k!$ where $k = \lambda_0 + \lambda_1 + \cdots + \lambda_{e-1}$.*

Proof Let us assume, by contradiction, that there exists a subset $S = \{x_1, \ldots, x_k\}$ of type $\boldsymbol{\lambda}$ in $\mathbb{Z}_m \rtimes_\varphi H \setminus \{id_{\mathbb{Z}_m \rtimes_\varphi H}\}$ that is not sequenceable and all the prime factors of m are larger than $k!$. Here S would be a set of type $\boldsymbol{\lambda}$ whose elements do not admit an ordering with different partial sums.

This implies that for any possible permutation $\omega \in \mathrm{Sym}(k)$, considered the ordering

$$\boldsymbol{x}_\omega = (x_{\omega(1)}, x_{\omega(2)}, \ldots, x_{\omega(k)}),$$

there exist h and t, with $h, t \in [0, k]$, $\{h, t\} \neq \{0, k\}$ and $h \neq t$, such that $s_h(\boldsymbol{x}_\omega) = s_t(\boldsymbol{x}_\omega)$, that is, assuming $h < t$,

$$x_{\omega(h+1)} \cdot x_{\omega(h+2)} \ldots x_{\omega(t-1)} \cdot x_{\omega(t)} = id. \tag{7}$$

Therefore we would have a solution $(\pi_{\mathbb{Z}_m}(x_1), \ldots, \pi_{\mathbb{Z}_m}(x_k))^T$ to the system of equations derived from (7) by considering the projection over $\mathbb{Z}_m$. In the following, for simplicity, we will denote $\pi_{\mathbb{Z}_m}(x_j)$ by x_j^1 and, using this notation, we are considering the system:

$$x^1_{\omega(h+1)} + \varphi_{\pi_H(x_{\omega(h+1)})} x^1_{\omega(h+2)} + \cdots + \varphi_{\pi_H(x_{\omega(h+1)}) \cdots \pi_H(x_{\omega(t-1)})} x^1_{\omega(t)} = 0 \qquad (8)$$

where, using the notation of the previous sections, and since φ satisfies property $(*)$, we set $\varphi(a)(x) = \varphi_a x$ for some constant $\varphi_a \in \{1, -1\}$. Denoted by M the $(k!) \times k$ matrix of the coefficients of this system, we note that, since φ satisfies property $(*)$, the coefficients of M belong to $\{-1, 0, 1\}$. It follows that the determinant of any square submatrix of M (that is big at most $k \times k$) is a sum of at most $k!$ terms that belong to ± 1. Therefore, if all the prime factors of m are larger than $k!$, a square submatrix M' of M is nonsingular in $\mathbb{Z}_m$ if and only if it is nonsingular in $\mathbb{Q}$.

Following the proof of Theorem 4.12 of [7] and assuming the existence a subset $S = \{x_1, \ldots, x_k\}$ of type $\boldsymbol{\lambda}$ in $\mathbb{Z}_m \rtimes_\varphi H \setminus \{id_{\mathbb{Z}_m \rtimes_\varphi H}\}$ that is not sequenceable, we find a subset of the same type in $\mathbb{Q} \rtimes_\varphi H \setminus \{id_{\mathbb{Q} \rtimes_\varphi H}\}$ that is also not sequenceable. However, because of Theorem 3 and since $\mathbb{Q}$ is a torsion-free abelian group, any subset $\widetilde{S}$ of type $\boldsymbol{\lambda}$ of $G \rtimes_\varphi H \setminus \{id_{G \rtimes_\varphi H}\}$ is sequenceable. But this is a contradiction and hence the thesis is verified. □

Corollary 3 *Let H be a finite group, let φ satisfy property $(*)$, k be a positive integer, and suppose that, for infinitely many odd primes p, any subset of size k of $\mathbb{Z}_p \rtimes_\varphi H \setminus \{id_{\mathbb{Z}_p \rtimes_\varphi H}\}$ is sequenceable. Then any subset of size k of $\mathbb{Z}_m \rtimes_\varphi H \setminus \{id_{\mathbb{Z}_m \rtimes_\varphi H}\}$ is sequenceable provided that the prime factors of m are all greater than $k!$.*

If we consider direct products of type $\mathbb{Z}_m \times H$, we can improve the result of Theorem 7. Indeed, with essentially the same proof of Theorem 22 of [7], we can state the following.

Theorem 8 *Let H be a finite group, k be a positive integer, and suppose that, for infinitely many odd primes p, any subset of size k of $\mathbb{Z}_p \times H \setminus \{id_{\mathbb{Z}_p \times H}\}$ is sequenceable. Then any subset of size k of $\mathbb{Z}_m \times H \setminus \{id_{\mathbb{Z}_m \times H}\}$ is sequenceable provided that the prime factors of m are all greater than $k!/2$.*

As regards dihedral groups, as a consequence of Theorem 2, we can prove Theorem 9.

Theorem 9 *Let m be such that all its prime factors are larger than $k!$. Then any subset S of D_{2m} such that $|S| \leq 12$ is sequenceable.*

Remark 7 Because of Remark 2, with essentially the same proof of Theorem 9, we can prove that if S is a subset of D_{2m} not of type $(k, 0)$, $k \leq 12$ and all the prime factors of m are larger than $k!$, then S admits a linear sequencing.

This implies that, under the hypothesis of Theorem 9, S admits a linear sequencing unless s_k is unavoidably the identity (s_k is the identity for any ordering).

4 Weak Sequenceability

In this section we consider again subsets S of a group $(G, \cdot)$ of the form $G = \mathbb{Z}_p \rtimes_\varphi H$ which is the semidirect product of $(\mathbb{Z}_p, +)$ and another group $(H, \cdot)$. As in the previous section we indicate by $x \cdot a$ (or by $y \cdot b$), where $x \in \mathbb{Z}_p$ and $a \in H$, a generic element of G. We investigate the t-weak sequenceability of subsets of semidirect products. Recall that in addition to requiring that $x_i - x_j \neq 0$ for $1 \leq i < j \leq k$ and $a_i = a_j$, we seek an ordering having no two of its partial sums s_i, s_j equal for $1 \leq i < j \leq k$ and $|i - j| \leq t$. Hence, modifying the expression of Eq. (2), we define, for $t < k$, the following polynomial

$$q_{\boldsymbol{a}} = \prod_{\substack{1 \leq i < j \leq k \\ a_i = a_j}} (x_j - x_i)$$
$$\times \prod_{\substack{0 \leq i < j \leq k \\ b_i = b_j \\ j - i \leq t \\ j \neq i+1}} (x_{i+1} + \varphi_{a_{i+1}} x_{i+2} + \varphi_{a_{i+1} \cdot a_{i+2}} x_{i+3} + \cdots + \varphi_{a_{i+1} \ldots a_{j-1}} x_j). \qquad (9)$$

In this case we have that a set $S = \{x_1 \cdot a_1, x_2 \cdot a_2, \ldots, x_k \cdot a_k\} \subseteq G \setminus \{id\}$ of size k is t-weak sequenceable if there exists an ordering $\boldsymbol{x_a} = (x_1 \cdot a_1, x_2 \cdot a_2, \ldots, x_k \cdot a_k)$ of its elements such that $q_{\boldsymbol{a}}(x_1, \ldots, x_k) \neq 0$.

Now, given a set $S = \{x_1 \cdot a_1, x_2 \cdot a_2, \ldots, x_k \cdot a_k\}$ of k elements, the idea is to fix, *a priori*, the first h elements $(x_1 \cdot a_1, x_2 \cdot a_2, \ldots, x_h \cdot a_h)$, where h is not too big, of the ordering in such a way that no of its partial sums s_i, s_j are equal when $|i - j| \leq t$ and $1 \leq i < j \leq h$. This can be expressed by requiring that $q_{\boldsymbol{a}'}(x_1, \cdots, x_h) \neq 0$ where $\boldsymbol{a}' = (a_1, \ldots, a_h)$ and we show that this can be done due the following Proposition whose proof is inspired by that of Proposition 2.2 of [4].

Proposition 3 *Let $S = \{x_1 \cdot a_1, x_2 \cdot a_2, \ldots, x_k \cdot a_k\} \subseteq G \setminus \{id\}$ be a set of size k and let h and t be positive integers such that $k - h \geq 2(t - 1)$ and $h \geq t - 1$. Denote by $\bar{a}$ the element that appears the most in $\pi_2(S)$, we assume that at least $2(t-1)$ elements of S have $\bar{a}$ as second coordinate. Then there exists an ordering of h-elements of S that we denote, up to relabeling, with $(x_1 \cdot a_1, x_2 \cdot a_2, \ldots, x_h \cdot a_h)$, such that*

(a) $q_{\boldsymbol{a}'}(x_1, \cdots, x_h) \neq 0$ *where* $\boldsymbol{a}' = (a_1, \ldots, a_h)$*;*
(b) $a_h, a_{h-1}, \ldots, a_{h-t+2}$ *are all equals to* $\bar{a}$*;*
(c) at least $t-1$ *elements of* $\{x_{h+1} \cdot a_{h+1}, \ldots, x_k \cdot a_k\}$ *have* $\bar{a}$ *as second coordinate.*

Proof Here we choose a set S' of $2(t-1)$ elements that have $\bar{a}$ as the second coordinate, and we set $S'' := S \setminus S'$. Now we prove that we can choose, inductively on ℓ, $(x_1 \cdot a_1, x_2 \cdot a_2, \dots, x_{h-t} \cdot a_{h-t})$ in S'' and $(x_{h-t+1} \cdot a_{h-t+1}, \dots, x_h \cdot a_h)$ in S' such that, for any $\ell \leq h$, we have $q_{(a_1,\dots,a_\ell)}(x_1, \cdots, x_\ell) \neq 0$. Then, properties (b) and (c) will be guaranteed by our choice for $x_1 \cdot a_1, \dots, x_h \cdot a_h$.

BASE CASE: Let $\ell = 1$. Since $q_{(a)}(x) = 1$ for any t and for any $x \cdot a$ of both S' and S'', the statement is realized for $\ell = 1$.

INDUCTIVE CASE: Let us assume the statement for $\ell \in \{1, \dots, m\}$ and let us prove it for $\ell = m+1$ where $m+1 \leq h$. Since the statement is true for $\ell = m$, there exists an m-tuple $(x_1 \cdot a_1, \dots, x_m \cdot a_m)$ such that

$$q_{(a_1,\dots,a_m)}(x_1, \cdots, x_m) \neq 0.$$

Given $x \cdot a \in G$, we set $s_m := (x_1 \cdot a_1) \cdots (x_m \cdot a_m) = y_m \cdot b_m$ and we define y and s such that $s = s_m \cdot (x \cdot a) = y \cdot b$ where $b = b_m \cdot a$. Then we have

$$\begin{aligned}
&\frac{q_{(a_1,\dots,a_m,a)}(x_1, \dots, x_m, x)}{q_{(a_1,\dots,a_m)}(x_1, \dots, x_m)} = \\
&= \prod_{1 \leq i < m+1,\ a=a_i} (x - x_i) \\
&\quad \times \prod_{\substack{0 \leq i < m \\ m+1-i \leq t,\ b_i = b}} (x_{i+1} + \varphi_{a_{i+1}} x_{i+2} + \varphi_{a_{i+1} \cdot a_{i+2}} x_{i+3} + \cdots + \varphi_{a_{i+1} \dots a_m} x).
\end{aligned}$$

Here, any element $x \cdot a$ of $S \setminus \{x_1 \cdot a_1, \dots, x_m \cdot a_m\}$ satisfies $\prod_{1 \leq i < m+1} (x - x_i) \neq 0$ if $a = a_i$. Hence, to have $\frac{q_{(a_1,\dots,a_m)}(x_1,\dots,x_m,x)}{q_{(a_1,\dots,a_m,a)}(x_1,\dots,x_m)} \neq 0$, it suffice to find x such that

$$\prod_{\max(0,m+1-t) \leq i < m,\ b_i = b} (x_{i+1} + \varphi_{a_{i+1}} x_{i+2} + \varphi_{a_{i+1} \cdot a_{i+2}} x_{i+3} + \cdots + \varphi_{a_{i+1} \dots a_m} x) \neq 0.$$

Note that there is at most one $x \cdot a \in S \setminus \{x_1 \cdot a_1, \dots, x_m \cdot a_m\}$ such that

$$(x_{i+1} + \varphi_{a_{i+1}} x_{i+2} + \varphi_{a_{i+1} \cdot a_{i+2}} x_{i+3} + \cdots + \varphi_{a_{i+1} \dots a_m} x) = 0$$

and $b_i = b$. Indeed this would imply that $s_i = s$ in G that is $(x_{i+1} \cdot a_{i+1}) \cdot \ldots \cdot (x_m \cdot a_m) \cdot (x \cdot a) = 0$ which is satisfied by exactly one element $x \cdot a$ of G. Since those relations are at most $t-1$, we have at most $t-1$ values $x \cdot a$ in $S \setminus \{x_1 \cdot a_1, \dots, x_m \cdot a_m\}$ such that

$$\prod_{\max(0,m+1-t) \leq i < m,\ b_i = b} (x_{i+1} + \varphi_{a_{i+1}} x_{i+2} + \varphi_{a_{i+1} \cdot a_{i+2}} x_{i+3} + \cdots + \varphi_{a_{i+1} \dots a_m} x) = 0.$$

Here we have that either $h - t + 1 > m$ or $h - t + 1 \leq m$. In the first case, we have that

$$|S'' \setminus \{x_1 \cdot a_1, \ldots, x_m \cdot a_m\}| = k - 2(t-1) - m \geq t + 1$$

where the inequality holds since $k - 2(t-1) - m = (k-m) - 2(t-1)$ and $k - m \geq k - h \geq 2(t-1)$. It follows that there exists $x \cdot a \in S'' \setminus \{x_1 \cdot a_1, \ldots, x_m \cdot a_m\}$ such that

$$\prod_{\max(0, m+1-t) \leq i < m,\ b_i = b} (x_{i+1} + \varphi_{a_{i+1}} x_{i+2} + \varphi_{a_{i+1} \cdot a_{i+2}} x_{i+3} + \cdots + \varphi_{a_{i+1} \ldots a_m} x) \neq 0.$$

In the second case, we have that

$$|S' \setminus \{x_1 \cdot a_1, \ldots, x_m \cdot a_m\}| = 2(t-1) - (m - (h - t + 1)) \geq t$$

where the last inequality holds since $m \leq h - 1$ and hence $2(t-1) - (m - (h - t + 1)) \geq 2(t-1) - (h - 1 - (h - t + 1)) = t$. It follows that also in this case, there exists $x \cdot a \in S' \setminus \{x_1 \cdot a_1, \ldots, x_m \cdot a_m\}$ such that

$$\prod_{\max(0, m+1-t) \leq i < m,\ b_i = b} (x_{i+1} + \varphi_{a_{i+1}} x_{i+2} + \varphi_{a_{i+1} \cdot a_{i+2}} x_{i+3} + \cdots + \varphi_{a_{i+1} \ldots a_m} x) \neq 0.$$

In both cases, we can find $x_{m+1} \cdot a_{m+1}$ that satisfies property (b) and (c) and which it is such that

$$\frac{q_{(a_1, \ldots, a_m, a_{m+1})}(x_1, \ldots, x_m, x_{m+1})}{q_{(a_1, \ldots, a_m)}(x_1, \ldots, x_m)} = \prod_{1 \leq i < m+1,\ a_{m+1} = a_i} (x_{m+1} - x_i) \cdot$$

$$\cdot \prod_{\max(0, m+1-t) \leq i < m,\ b_i = b_{m+1}}$$

$$\times (x_{i+1} + \varphi_{a_{i+1}} x_{i+2} + \varphi_{a_{i+1} \cdot a_{i+2}} x_{i+3} + \cdots + \varphi_{a_{i+1} \ldots a_m} x_{m+1}) \neq 0.$$

Since $\mathbb{Z}_p$ is a field and due to the inductive hypothesis $q_{(a_1, \ldots, a_m, a)}(x_1, \ldots, x_m) \neq 0$, we also have that

$$\frac{q_{(a_1, \ldots, a_m, a_{m+1})}(x_1, \ldots, x_m, x_{m+1})}{q_{(a_1, \ldots, a_m)}(x_1, \ldots, x_m)} \cdot$$

$$q_{(a_1, \ldots, a_m)}(x_1, \ldots, x_m) = q_{(a_1, \ldots, a_m, a_{m+1})}(x_1, \ldots, x_{m+1}) \neq 0 \tag{10}$$

which completes the proof. □

Remark 8 Note that the hypothesis that at least $2(t-1)$ elements of S has the same second coordinate is always satisfied when

$$k > (2t-3)\frac{|G|}{p} = (2t-3)|H|.$$

This means that the approach of Proposition 3 can be always tried assuming that k is big enough. For smaller values of k one could fix the first elements differently, as done for example in Proposition 2.2 of [4], but, since then one should consider a huge number of cases (see also the following Remark 9), we consider explicitly only this asymptotic case.

In the following we assume that we have fixed, according to Proposition 3,

$$\{x_1 \cdot a_1, x_2 \cdot a_2, \ldots, x_h \cdot a_h\} \subseteq S$$

that satisfies properties (a), (b) and (c) of such proposition. We note that every $x \cdot a \in S \setminus \{x_1 \cdot a_1, x_2 \cdot a_2, \ldots, x_h \cdot a_h\}$ is such that $x - x_i \neq 0$ for any $i \in \{1, \ldots, h\}$ such that $a = a_i$. Therefore, if $\boldsymbol{a}$ is a quotient sequencing that extends $\boldsymbol{a}''$, it is left to find a nonzero point for the polynomial

$$\frac{q_{\boldsymbol{a}}(x_1, \ldots, x_h, x_{h+1}, \ldots, x_k)}{q_{\boldsymbol{a}'}(x_1, \cdots, x_h)\prod_{1\leq i\leq h<j\leq k,\ a_j=a_i}(x_j - x_i)}.$$

Since the free variables are now $x_{h+1}, \ldots, x_k$, we set $\ell := k - h$ and $z_i := x_{i+h}$; here the constrain $k - h \geq 2(t-1)$ of Proposition 3 becomes $\ell \geq 2(t-1)$. Then we denote by $h_{\boldsymbol{a},\ell}$ the polynomial

$$h_{\boldsymbol{a},\ell}(z_1, \ldots, z_\ell) := \frac{q_{\boldsymbol{a}}(x_1, \ldots, x_h, z_1, \ldots, z_\ell)}{q_{\boldsymbol{a}'}(x_1, \cdots, x_h)\prod_{1\leq i\leq 1<j\leq \ell,\ a_j=a_i}(z_j - x_i)}. \tag{11}$$

Since we are also assuming that $k - \ell \geq t-1$, that is, $k-(t-1) \geq \ell \geq 2(t-1)$, denoting by $\boldsymbol{a}''$ the quotient sequencing given by the last ℓ elements of $\boldsymbol{a}$, we obtain the following expression

$$\begin{aligned} h_{\boldsymbol{a},\ell}(z_1, \cdots, z_\ell) = q_{\boldsymbol{a}''}(z_1, \cdots, z_\ell)\cdot \\ \prod_{\substack{0\leq i\leq t-1;\\ 1\leq j\leq t-i-1;\\ b_{k-\ell-i-1}=b_{(k-\ell)+j}}} (x_{k-\ell-i} + \varphi_{\bar{a}}x_{k-\ell-i+1} + \cdots + \varphi_{\bar{a}^i}x_{k-\ell} + \\ \varphi_{\bar{a}^{i+1}}z_1 + \cdots + \varphi_{\bar{a}^{i+1}\cdot a_{k-\ell+1}\cdot\cdots\cdot a_{k-\ell+j-1}}z_j). \end{aligned} \tag{12}$$

Now we aim to apply the Non-Vanishing Corollary (of the Combinatorial Nullstellensatz) to the polynomial $h_{\boldsymbol{a},\ell}$. For this purpose, it is enough to consider the terms of $h_{\boldsymbol{a},\ell}$ of maximal degree in the variables $z_1, \ldots, z_\ell$ that are the ones where no x_i appears. We denote by $r_{\boldsymbol{a},\ell}$ the polynomial given by those terms, that is

$$r_{\boldsymbol{a},\ell} := q_{\boldsymbol{a}''}(z_1, \cdots, z_\ell) \cdot \prod_{\substack{0 \leq i \leq t-1; \\ 1 \leq j \leq t-i-1; \\ b_{k-\ell-i-1} = b_{(k-\ell)+j}}} (\varphi_{\bar{a}^{i+1}} z_1 + \varphi_{\bar{a}^{i+1} \cdot a_{k-\ell+1}} z_2 + \cdots + \varphi_{\bar{a}^{i+1} \cdot a_{k-\ell+1} \cdot \cdots \cdot a_{k-\ell+j-1}} z_j). \tag{13}$$

Since $b_{k-\ell-i-1} = b_{(k-\ell)+j}$ if and only if

$$a_{k-\ell-i} \cdot a_{k-\ell-1} \cdot \ldots \cdot a_{k-\ell} \cdot a_{k-\ell+1} \cdot \ldots \cdot a_{k-\ell+j} = id$$

and $a_{k-\ell-i} = a_{k-\ell-1} = \ldots = a_{k-\ell} = \bar{a}$, we can state the following, simple but very powerful, remark.

Remark 9 Since we are assuming that $a_h, a_{h-1}, \ldots, a_{h-t+1}$ are all equals to $\bar{a}$, the expression of $r_{\boldsymbol{a},\ell}$ does not depend on all the quotient sequence $\boldsymbol{a}$ but only by $\boldsymbol{a}''$ and $\bar{a}$. Here property b of Proposition 3 is crucial since, otherwise, given $\boldsymbol{a}''$, we would have $|H|^{t-1}$ possibilities for the polynomial $r_{\boldsymbol{a},\ell}(z_1, \cdots, z_\ell)$ according to the sequence $a_h, a_{h-1}, \ldots, a_{h-t+1}$.

In the following, we just denote this polynomial by $r_{\boldsymbol{a}'',\bar{a},\ell}$.

Indeed Remark 9 means that, after these manipulations, we are left to consider a polynomial that does not depend on $k = |S|$, and hence we have chances to get a result that is very general on k.

To apply the Non-Vanishing Corollary, we also need to find, for any quotient sequencing $\boldsymbol{a}''$ of length ℓ, a nonzero coefficient in $r_{\boldsymbol{a}'',\bar{a},\ell}$ that divides the bounding monomial. Moreover, due to property c of Proposition 3, we may assume that $\bar{a}$ appears at least $t-1$ times in $\boldsymbol{a}''$.

4.1 Computational Results

As done for the sequenceability problem in the more general setting, we consider here groups of type D_{2p} and G_{3p}. Also, since the case $G = \mathbb{Z}_p \times \mathbb{Z}_e$ has not been considered in [4], we consider these groups here.

Example 3 Example Let $t = 4$, $\ell = 2(t-1) = 6$ and $\bar{a} = 0$. Let p be a prime and suppose $S \subseteq D_{2p} \setminus \{id\}$ with $|S| = k \geq 11$ and suppose that S has at least 6 elements with coset $\bar{a}$. Let $S' = S \setminus \{x_1 \cdot a_1, \ldots, x_{k-\ell} \cdot a_{k-\ell}\}$ be of type (3, 3). The sequence $\boldsymbol{a}'' = (1, 0, 1, 0, 1, 0)$ has partial sums (0, 1, 1, 0, 0, 1, 1).

By Proposition 3, we desire a sequencing of S' of the form

$$(z_1 \cdot 1, z_2 \cdot 0, z_3 \cdot 1, z_4 \cdot 0, z_5 \cdot 1, z_6 \cdot 0)\,.$$

The polynomial we need to study is the one defined in (13). Hence we get

$$\begin{aligned} r_{\boldsymbol{a}'',\bar{a},\ell} &= (z_1 - z_2 - z_3)(z_3 - z_1)(z_5 - z_1)(z_4 - z_2)(z_6 - z_2)(z_5 - z_3)(z_6 - z_4) \\ &\quad \times (z_1 - z_2 - z_3)(z_1 - z_2 - z_3 + z_4)(z_2 + z_3 - z_4 - z_5) \\ &\quad \times (z_3 - z_4 - z_5)(z_3 - z_4 - z_5 + z_6) \end{aligned}$$

To apply the Non-Vanishing Corollary we need a monomial of this polynomial which divides the bounding monomial $z_1^2 z_2^2 z_3^2 z_4^2 z_5^2 z_6^2$ with a nonzero coefficient. One such is $z_1^2 z_2^2 z_3^2 z_4^2 z_5^2 z_6^2$, which has coefficient -12. Hence whenever S has one of these types it is t-weakly sequenceable.

We are now ready to prove the main result of this section.

Theorem 10 *Let $n = pe$ with $p > 3$ prime and let G be a group of size n. Then subsets S of size k of G are t-weakly sequenceable whenever $e \in \{1, 2, 3\}$, k is large enough and $t \le 6$.*

Table 3 Monomials and their coefficients sufficient for the proof of Theorem 10 in the case $t = 6$ and $G = \mathbb{Z}_p \times \mathbb{Z}_2$

$\bar{a}$	λ	$\mathbf{a}''$	Deg	Monomial/s	Coefficient/s
0	(5, 5)	(0, 1, 0, 0, 1, 1, 1, 0, 0, 1)	40	$x_1^4 x_2^4 x_3^4 x_4^4 x_5^4 x_6^4 x_7^4 x_8^4 x_9^4 x_{10}^4$	$-5 \cdot 467$
0	(6, 4)	(1, 0, 0, 0, 1, 0, 1, 0, 1, 0)	38	$x_2^4 x_3^5 x_4^5 x_5^3 x_6^5 x_7^3 x_8^5 x_9^3 x_{10}^5$	-1063
				$x_1 x_2^3 x_3^5 x_4^5 x_5^3 x_6^5 x_7^3 x_8^5 x_9^3 x_{10}^5$	$2^4 \cdot 503$
0	(7, 3)	(1, 0, 0, 0, 1, 0, 0, 0, 1, 0)	36	$x_3^2 x_4^6 x_5^2 x_6^6 x_7^6 x_8^6 x_9^2 x_{10}^6$	2
0	(8, 2)	(1, 0, 0, 0, 1, 0, 0, 0, 0, 0)	45	$x_3^2 x_4^7 x_5 x_6^7 x_7^7 x_8^7 x_9^7 x_{10}^7$	-2^2
0	(9, 1)	(0, 0, 0, 0, 1, 0, 0, 0, 0, 0)	66	$x_1^5 x_2^7 x_3^7 x_4^8 x_6^7 x_7^8 x_8^8 x_9^8 x_{10}^8$	$-2^5 \cdot 3$
1	(4, 6)	(0, 1, 1, 1, 0, 0, 1, 0, 1, 1)	42	$x_1^3 x_2^5 x_3^5 x_4^5 x_5^3 x_6^3 x_7^5 x_8^3 x_9^5 x_{10}^5$	-1187
1	(3, 7)	(1, 0, 1, 1, 1, 1, 1, 0, 1, 0)	47	$x_1^5 x_2^2 x_3^6 x_4^6 x_5^6 x_6^6 x_7^6 x_8^2 x_9^6 x_{10}^2$	$2 \cdot 3^2 \cdot 1511$
				$x_1^6 x_2 x_3^6 x_4^6 x_5^6 x_6^6 x_7^6 x_8^2 x_9^6 x_{10}^2$	$23 \cdot 499$
1	(2, 8)	(1, 0, 1, 1, 1, 1, 1, 1, 1, 0)	54	$x_1^4 x_2 x_3^6 x_4^7 x_5^7 x_6^7 x_7^7 x_8^7 x_9^7 x_{10}$	$-2^5 \cdot 5^2$
				$x_1^5 x_2 x_3^5 x_4^7 x_5^7 x_6^7 x_7^7 x_8^7 x_9^7 x_{10}$	-4337
1	(1, 9)	(1, 0, 1, 1, 1, 1, 1, 1, 1, 1)	62	$x_1^5 x_3^3 x_4^6 x_5^8 x_6^8 x_7^8 x_8^8 x_9^8 x_{10}^8$	-2
1	(0, 10)	(1, 1, 1, 1, 1, 1, 1, 1, 1, 1)	75	$x_1^3 x_2^4 x_3^5 x_4^9 x_5^9 x_6^9 x_7^9 x_8^9 x_9^9 x_{10}^9$	$-2^4 \cdot 5^2$
				$x_1^3 x_2^5 x_3^5 x_4^8 x_5^9 x_6^9 x_7^9 x_8^9 x_9^9 x_{10}^9$	$-7 \cdot 853$

Proof The case $e = 1$ follows from the results of [4].

As already noted in the proof of Theorem 2, a group of size $2p$ is either $\mathbb{Z}_p \times \mathbb{Z}_2$ or the dihedral group D_{2p} and a group of size $3p$ is either $\mathbb{Z}_p \times \mathbb{Z}_3$ or G_{3p} for some value of r. Here, due to Remark 8, if $k \geq (2t-3)e+1$, S admits a coset whose size is at least $2(t-1)$ and hence, due to Proposition 3, we can apply Non-Vanishing Corollary to the polynomial of Eq. (13) for suitable quotient sequences and nonzero monomial (see Tables 3, 4, 5 and 6 below). These results are obtained using SageMath [19].

Table 4 Monomials and their coefficients sufficient for the proof of Theorem 10 in the case $t = 6$ and $G = \mathbb{Z}_p \times \mathbb{Z}_3$

$\bar{a}$	$\boldsymbol{\lambda}$	$\mathbf{a}''$	Deg	Monomial/s	Coefficient/s
0	(5, 3, 2)	(1, 0, 0, 1, 0, 0, 0, 2, 1, 2)	25	$x_1^2x_2x_3^4x_4^2x_5^4x_6^4x_7^4x_8x_9^2x_{10}$	$2 \cdot 3^2$
0	(5, 4, 1)	(1, 0, 0, 1, 0, 0, 0, 1, 1, 2)	22	$x_3^4x_4x_5^3x_6^4x_7^4x_8^3x_9^3$	1
0	(5, 5, 0)	(0, 1, 0, 1, 0, 0, 0, 1, 1, 1)	33	$x_1^2x_2x_3^4x_4^4x_5^4x_6^4x_7^4x_8^2x_9^4x_{10}^4$	$3 \cdot 5^2$
0	(6, 2, 2)	(1, 0, 0, 1, 0, 0, 0, 0, 2, 2)	25	$x_3^3x_4x_5^4x_6^5x_7^5x_8^5x_9x_{10}$	-1
0	(6, 3, 1)	(1, 0, 0, 1, 0, 0, 0, 0, 1, 2)	30	$x_2x_3^5x_4^2x_5^5x_6^5x_7^5x_8^5x_9^2$	-2^4
0	(6, 4, 0)	(1, 0, 0, 1, 0, 0, 0, 0, 1, 1)	28	$x_3^3x_4x_5^4x_6^5x_7^5x_8^5x_9^2x_{10}^3$	-1
0	(7, 2, 1)	(1, 0, 0, 1, 0, 0, 0, 0, 0, 2)	33	$x_3^3x_4x_5^5x_6^6x_7^6x_8^6x_9^6$	-2^2
0	(7, 3, 0)	(1, 0, 0, 1, 0, 0, 0, 0, 0, 1)	35	$x_3^3x_4x_5^5x_6^6x_7^6x_8^6x_9^6x_{10}^2$	-2^2
0	(8, 1, 1)	(1, 0, 0, 0, 2, 0, 0, 0, 0, 0)	44	$x_3^2x_4^7x_6^7x_7^7x_8^7x_9^7x_{10}^7$	-2^2
0	(8, 2, 0)	(1, 0, 0, 0, 1, 0, 0, 0, 0, 0)	42	$x_3^2x_4^5x_5x_6^6x_7^7x_8^7x_9^7x_{10}^7$	-2^2
0	(9, 1, 0)	(0, 0, 0, 1, 0, 0, 0, 0, 0, 0)	66	$x_1^5x_2^6x_3^8x_5^7x_6^8x_7^8x_8^8x_9^8x_{10}^8$	$2^2 \cdot 7$
1	(3, 5, 2)	(0, 1, 1, 0, 1, 1, 1, 2, 0, 2)	27	$x_1x_2^4x_3^4x_4^2x_5^4x_6^4x_7^4x_8x_9^2x_{10}$	-73
				$x_1^2x_2^3x_3^4x_4^2x_5^4x_6^4x_7^4x_8x_9^2x_{10}$	-233
1	(2, 5, 3)	(0, 1, 0, 1, 2, 1, 1, 1, 2, 2)	28	$x_1x_2^4x_3x_4^4x_5^2x_6^4x_7^4x_8^4x_9^2x_{10}^2$	7
1	(4, 5, 1)	(0, 1, 1, 0, 1, 1, 1, 0, 0, 2)	31	$x_1^2x_2^4x_3^4x_4^3x_5^4x_6^4x_7^4x_8^3x_9^3$	$-5 \cdot 43$
				$x_1^3x_2^3x_3^4x_4^3x_5^4x_6^4x_7^4x_8^3x_9^3$	$-3 \cdot 13 \cdot 17$
1	(1, 5, 4)	(1, 2, 2, 0, 2, 2, 1, 1, 1, 1)	32	$x_1^4x_2^3x_3^3x_5^3x_6^3x_7^4x_8^4x_9^4x_{10}^4$	-7
1	(0, 5, 5)	(1, 2, 1, 2, 1, 1, 1, 2, 2, 2)	38	$x_1^3x_2^4x_3^4x_4^4x_5^4x_6^3x_7^4x_8^4x_9^4x_{10}^4$	$-2 \cdot 3 \cdot 5$
1	(2, 6, 2)	(0, 1, 0, 1, 1, 2, 2, 1, 1, 1)	30	$x_1x_2^2x_3x_4^4x_5^5x_6x_7x_8^5x_9^5x_{10}^5$	-5
1	(3, 6, 1)	(0, 1, 0, 1, 1, 0, 2, 1, 1, 1)	32	$x_1x_2^2x_3^2x_4^5x_5^5x_6^2x_8^5x_9^5x_{10}^5$	2^2
1	(1, 6, 3)	(0, 1, 0, 1, 1, 0, 2, 1, 1, 1)	32	$x_1x_2^2x_3^2x_4^5x_5^5x_6^2x_8^5x_9^5x_{10}^5$	2^2
1	(4, 6, 0)	(0, 1, 1, 0, 1, 1, 1, 1, 0, 0)	34	$x_1x_2x_3^3x_4^3x_5^5x_6^5x_7^5x_8^5x_9^3x_{10}^3$	$-2 \cdot 3$

(continued)

Table 6 Monomials and their coefficients sufficient for the proof of Theorem 10 in the case $t = 6$ and $G = G_{3p}$

$\bar{a}$	$\boldsymbol{\lambda}$	$\mathbf{a}''$	Deg	Monomial/s	Coefficient/s
0	(5, 3, 2)	(1, 0, 0, 1, 0, 0, 0, 2, 1, 2)	25	$x_1^2x_2^4x_3^4x_4^2x_5^2x_6^4x_7^4x_9^2x_{10}$	$-r-1$
0	(5, 4, 1)	(1, 0, 0, 1, 0, 0, 0, 1, 1, 2)	22	$x_1^3x_2^4x_3^3x_4^3x_5x_6^2x_7^4x_8x_9$	1
0	(5, 5, 0)	(0, 1, 0, 1, 0, 0, 0, 1, 1, 1)	33	$x_1^2x_2^2x_3^2x_4^3x_5^4x_6^4x_7^4x_8^4x_9^4x_{10}^4$	$124r+294$
0	(6, 2, 2)	(1, 0, 0, 1, 0, 0, 0, 0, 2, 2)	25	$x_1x_2^5x_3^2x_4x_5^3x_6^2x_7^5x_8^5x_{10}$	1
0	(6, 3, 1)	(1, 0, 0, 1, 0, 0, 0, 0, 1, 2)	30	$x_1^2x_2^3x_3^4x_4x_5^5x_6^5x_7^4x_8^4x_9^2$	1
0	(6, 4, 0)	(1, 0, 0, 1, 0, 0, 0, 0, 1, 1)	28	$x_2^2x_3^3x_4x_5^4x_6^4x_7^5x_8^4x_9^2x_{10}^3$	2
0	(7, 2, 1)	(1, 0, 0, 1, 0, 0, 0, 0, 0, 2)	33	$x_3^3x_4x_5^6x_6^5x_7^6x_8^6x_9^6$	2
0	(7, 3, 0)	(1, 0, 0, 1, 0, 0, 0, 0, 0, 1)	35	$x_3^3x_4x_5^6x_6^6x_7^6x_8^5x_9^6x_{10}^2$	2
0	(8, 1, 1)	(1, 0, 0, 0, 2, 0, 0, 0, 0, 0)	44	$x_2^3x_3^4x_4^2x_6^7x_7^7x_8^7x_9^7x_{10}^7$	$12r+12$
0	(8, 2, 0)	(1, 0, 0, 0, 1, 0, 0, 0, 0, 0)	42	$x_2^3x_3^4x_4^4x_5x_6^5x_7^5x_8^7x_9^7x_{10}^6$	5
0	(9, 1, 0)	(0, 0, 0, 1, 0, 0, 0, 0, 0, 0)	66	$x_1^5x_2^7x_3^7x_5^8x_6^8x_7^7x_8^8x_9^8x_{10}^8$	$2^2 \cdot 5^2$
1	(3, 5, 2)	(0, 1, 1, 0, 1, 1, 1, 2, 0, 2)	27	$x_1x_2^4x_3^4x_4^2x_5^4x_6^4x_7^4x_8x_9^2x_{10}$	$-3055r-1872$
1	(2, 5, 3)	(0, 1, 1, 0, 1, 1, 1, 2, 2, 2)	28	$x_1x_2^4x_3^4x_4x_5^4x_6^4x_7^4x_8^2x_9^2x_{10}^2$	$3993r+2619$
1	(4, 5, 1)	(0, 1, 1, 0, 1, 1, 1, 0, 0, 2)	31	$x_1^2x_2^4x_3^4x_4^3x_5^4x_6^4x_7^4x_8^3x_9^3$	$3130r+21$
1	(1, 5, 4)	(1, 2, 2, 1, 2, 2, 1, 1, 1, 0)	32	$x_1^4x_2^3x_3^3x_4^4x_5^3x_6^3x_7^4x_8^4x_9^4$	$-4043r-1625$
1	(0, 5, 5)	(1, 2, 1, 2, 1, 1, 1, 2, 2, 2)	38	$x_1^3x_2^3x_3^4x_4^4x_5^4x_6^4x_7^4x_8^4x_9^4x_{10}^4$	$7128r-12852$
1	(2, 6, 2)	(0, 1, 0, 1, 1, 2, 2, 1, 1, 1)	30	$x_1x_2^3x_3x_4^3x_5^5x_6x_7x_8^5x_9^5x_{10}^5$	$-424r-595$
1	(3, 6, 1)	(0, 1, 0, 1, 1, 0, 2, 1, 1, 1)	36	$x_1^2x_2^5x_3^2x_4^5x_5^5x_6^2x_8^5x_9^5x_{10}^5$	$-566r+1653$
1	(1, 6, 3)	(1, 0, 1, 1, 2, 2, 2, 1, 1, 1)	34	$x_1^4x_3^4x_4^5x_5^2x_6^2x_7^2x_8^5x_9^5x_{10}^5$	$61364r+25951$
1	(4, 6, 0)	(0, 1, 1, 0, 1, 1, 1, 1, 0, 0)	34	$x_1^3x_2^3x_3^3x_4^3x_5^3x_6^3x_7^5x_8^5x_9^3x_{10}^3$	$-1272r+653$
1	(0, 6, 4)	(2, 1, 1, 2, 1, 1, 1, 1, 2, 2)	37	$x_1^3x_2^3x_3^3x_4^3x_5^4x_6^5x_7^5x_8^5x_9^3x_{10}^3$	$-5321r-5949$
1	(2, 7, 1)	(0, 1, 1, 0, 1, 1, 1, 1, 1, 2)	35	$x_1x_2^3x_3^3x_4x_5^3x_6^6x_7^6x_8^6x_9^6$	$-291r+1$
1	(1, 7, 2)	(1, 0, 1, 2, 1, 1, 1, 1, 1, 2)	40	$x_1^4x_3^4x_4x_5^6x_6^6x_7^6x_8^6x_9^6x_{10}$	$11711r+1287$
1	(3, 7, 0)	(0, 1, 1, 0, 1, 1, 1, 1, 1, 0)	36	$x_1^2x_2^3x_3^3x_4^2x_5^3x_6^3x_7^6x_8^6x_9^6x_{10}^2$	$5839r+1451$
1	(0, 7, 3)	(2, 1, 1, 2, 1, 1, 1, 1, 1, 2)	40	$x_1^2x_2^3x_3^3x_4^2x_5^4x_6^6x_7^6x_8^6x_9^6x_{10}^2$	$-1365r-1577$
1	(1, 8, 1)	(1, 0, 1, 1, 1, 1, 1, 1, 1, 2)	43	$x_1^4x_3^4x_4^4x_5^4x_6^6x_7^7x_8^7x_9^7$	$239r-1811$
1	(2, 8, 0)	(1, 0, 1, 1, 1, 1, 1, 1, 1, 0)	43	$x_1^4x_2x_3^4x_4^4x_5^4x_6^4x_7^7x_8^7x_9^7x_{10}$	$-10919r-9004$
1	(0, 8, 2)	(1, 2, 1, 1, 1, 1, 1, 1, 1, 2)	47	$x_1^4x_2x_3^4x_4^4x_5^5x_6^7x_7^7x_8^7x_9^7x_{10}$	$5278r+39848$
1	(1, 9, 0)	(1, 0, 1, 1, 1, 1, 1, 1, 1, 1)	51	$x_1^4x_3^4x_4^4x_5^5x_6^5x_7^5x_8^8x_9^8x_{10}^8$	$-12212r-8422$
1	(0, 9, 1)	(1, 2, 1, 1, 1, 1, 1, 1, 1, 1)	54	$x_1^4x_3^4x_4^4x_5^6x_6^6x_7^6x_8^8x_9^8x_{10}^8$	$-82561r-26061$
1	(0, 10, 0)	(1, 1, 1, 1, 1, 1, 1, 1, 1, 1)	65	$x_1^2x_2^4x_3^4x_4^4x_5^6x_6^9x_7^9x_8^9x_9^9x_{10}^9$	$7028r+8686$
2	(3, 2, 5)	(0, 2, 2, 0, 2, 2, 2, 1, 0, 1)	27	$x_1x_2^4x_3^4x_4^2x_5^4x_6^4x_7^4x_8x_9^2x_{10}$	$3055r+1183$
2	(2, 3, 5)	(0, 2, 2, 0, 2, 2, 2, 1, 1, 1)	28	$x_1x_2^4x_3^4x_4x_5^4x_6^4x_7^4x_8^2x_9^2x_{10}^2$	$-3993r-1374$
2	(4, 1, 5)	(0, 2, 2, 0, 2, 2, 2, 0, 0, 1)	31	$x_1^2x_2^4x_3^4x_4^3x_5^4x_6^4x_7^4x_8^3x_9^3$	$-3130r-3109$

(continued)

Table 6 (continued)

$\bar{a}$	λ	$\mathbf{a}''$	Deg	Monomial/s	Coefficient/s
2	(1, 4, 5)	(2, 1, 1, 2, 1, 1, 2, 2, 2, 0)	32	$x_1^4x_2^3x_3^3x_4^4x_5^3x_6^3x_7^4x_8^4x_9^4$	$4043r + 2418$
2	(5, 0, 5)	(0, 2, 0, 2, 0, 0, 0, 2, 2, 2)	33	$x_1^2x_2^2x_3^2x_4^3x_5^4x_6^4x_7^4x_8^4x_9^4x_{10}^4$	$-124r + 170$
2	(2, 2, 6)	(0, 2, 0, 2, 2, 1, 1, 2, 2, 2)	30	$x_1x_2^3x_3x_4^3x_5^5x_6x_7x_8^5x_9^5x_{10}^5$	$424r - 171$
2	(3, 1, 6)	(0, 2, 0, 2, 2, 0, 1, 2, 2, 2)	36	$x_1^2x_2^5x_3^2x_4^5x_5^5x_6^2x_8^5x_9^5x_{10}^5$	$566r + 2219$
2	(1, 3, 6)	(2, 0, 2, 2, 1, 1, 1, 2, 2, 2)	34	$x_1^4x_3^4x_4^5x_5^2x_6^2x_7^2x_8^5x_9^5x_{10}^5$	$-61364r - 35413$
2	(4, 0, 6)	(0, 2, 2, 0, 2, 2, 2, 2, 0, 0)	34	$x_1^3x_2^3x_3^3x_4^3x_5^3x_6^3x_7^5x_8^5x_9^3x_{10}^3$	$1272r + 1925$
2	(0, 4, 6)	(1, 2, 2, 1, 2, 2, 2, 2, 1, 1)	37	$x_1^3x_2^3x_3^3x_4^3x_5^4x_6^5x_7^5x_8^5x_9^3x_{10}^3$	$5321r - 628$
2	(2, 1, 7)	(0, 2, 2, 0, 2, 2, 2, 2, 2, 1)	35	$x_1x_2^3x_3^3x_4x_5^3x_6^6x_7^6x_8^6x_9^6$	$291r + 29$
2	(1, 2, 7)	(2, 0, 2, 1, 2, 2, 2, 2, 2, 1)	40	$x_1^4x_3^4x_4x_5^6x_6^6x_7^6x_8^6x_9^6x_{10}$	$-11711r - 10424$
2	(3, 0, 7)	(0, 2, 2, 0, 2, 2, 2, 2, 2, 0)	36	$x_1^2x_2^3x_3^3x_4^2x_5^3x_6^3x_7^6x_8^6x_9^6x_{10}^2$	$-5839r - 4388$
2	(0, 3, 7)	(1, 2, 2, 1, 2, 2, 2, 2, 2, 1)	40	$x_1^2x_2^3x_3^3x_4^2x_5^4x_6^6x_7^6x_8^6x_9^6x_{10}^2$	$1365r - 212$
2	(1, 1, 8)	(2, 0, 2, 2, 2, 2, 2, 2, 2, 1)	43	$x_1^4x_3^4x_4^4x_5^4x_6^6x_7^7x_8^7x_9^7$	$-239r - 2050$
2	(2, 0, 8)	(2, 0, 2, 2, 2, 2, 2, 2, 2, 0)	43	$x_1^4x_2x_3^4x_4^4x_5^4x_6^4x_7^7x_8^7x_9^7x_{10}$	$10919r + 1915$
2	(0, 2, 8)	(2, 1, 2, 2, 2, 2, 2, 2, 2, 1)	47	$x_1^4x_2x_3^4x_4^4x_5^5x_6^7x_7^7x_8^7x_9^7x_{10}$	$-5278r + 34570$
2	(1, 0, 9)	(2, 0, 2, 2, 2, 2, 2, 2, 2, 2)	51	$x_1^4x_3^4x_4^4x_5^5x_6^5x_7^5x_8^8x_9^8x_{10}^8$	$12212r + 3790$
2	(0, 1, 9)	(2, 1, 2, 2, 2, 2, 2, 2, 2, 2)	54	$x_1^4x_3^4x_4^4x_5^6x_6^6x_7^6x_8^8x_9^8x_{10}^8$	$82561r + 56500$
2	(0, 0, 10)	(2, 2, 2, 2, 2, 2, 2, 2, 2, 2)	65	$x_1^2x_2^4x_3^4x_4^4x_5^6x_6^9x_7^9x_8^9x_9^9x_{10}^9$	$-7028r + 1658$

References

1. N. Alon, Combinatorial Nullstellensatz. Combin. Probab. Comput. **8**, 7–29 (1999)
2. B. Alspach, G. Liversidge, On strongly sequenceable abelian groups. Art Discrete Appl. Math. **3**, 19pp. (2020)
3. B.A. Anderson, E.C. Ihrig, Every finite solvable group with a unique element of order two, except the quaternion group, has a symmetric sequencing. J. Combin. Des. **1**, 3–14 (1993)
4. S. Costa, S. Della Fiore, Weak sequenceability in cyclic groups. J. Combin. Des. **30**, 735–751 (2022)
5. S. Costa, M.A. Pellegrini, Some new results about a conjecture by Brian Alspach. Arch. Math. **115**, 479–488 (2020)
6. S. Costa, F. Morini, A. Pasotti, M.A. Pellegrini, A problem on partial sums in abelian groups. Discrete Math. **341**, 705–712 (2018)
7. S. Costa, S. Della Fiore, M.A. Ollis, S. Z. Rovner-Frydman, On sequences in cyclic groups with distinct partial sums. Electron. J. Combin. **29**, #P3.33 (2022)
8. B. Gordon, Sequences in groups with distinct partial products. Pac. J. Math. **11**, 1309–1313 (1961)
9. R.L. Graham, On sums of integers taken from a fixed sequence, in *Proceedings, Washington State University Conference on Number Theory* (1971), pp. 22–40
10. J. Hicks, M.A. Ollis, J.R. Schmitt, Distinct partial sums in cyclic groups: polynomial method and constructive approaches. J. Combin. Des. **27**, 369–385 (2019)
11. J. Isbell, Sequencing certain dihedral groups. Discrete Math. **85**, 323–328 (1990)

12. A.D. Keedwell, On the sequenceability of non-abelian groups of order pq. Discrete Math. **37**, 203–216 (1981)
13. P. Li, Sequencing the dihedral groups D_{4k}. Discrete Math. **175**, 271–276 (1997)
14. M. Michalek, A short proof of Combinatorial Nullstellensatz. Am. Math. Monthly **117**, 821–823 (2010)
15. M.A. Ollis, Terraces for small groups. J. Combin. Math. Comput. **108**, 231–244 (2019)
16. M.A. Ollis, Sequences in dihedral groups with distinct partial products. Australas. J. Combin. **78**, 35–60 (2020)
17. M.A. Ollis, C.R. Tripp, The spectrum of group-based complete latin squares. Electron. J. Combin. **26** #3.15, 13pp. (2019)
18. A. Pasotti, J.H. Dinitz, A survey of Heffter arrays. Fields Inst. Commun. **86**, 353–392 (2024)
19. Sage Developers, Sage Mathematics Software (Version 9.5.0) (The SAGE Group, 2022). http://www.sagemath.org
20. W. Sawin, comment on the post "Ordering subsets of the cyclic group to give distinct partial sums" (MathOverflow, 2015). https://mathoverflow.net/q/202857

Type (a, b, c) Face-magic Labelings of Calendula Graphs

Bryan Freyberg, Gretta Jensen, and Kaysie Scott

Abstract The calendula graph $Cal(m, n)$ is a planar graph consisting of a single central m-cycle and m edge-disjoint peripheral n-cycles, each of which shares exactly one edge with the central m-cycle. Let $G = (V, E, F)$ be a planar graph and $a, b, c \in \{0, 1\}$. A face-magic labeling of type (a, b, c) is an assignment of a labels to each vertex, b labels to each edge, and c labels to each face from the set of integer labels $\{1, 2, \ldots, a|V| + b|E| + c|F|\}$ such that each label is used exactly once, and for each s-sided face f, the sum of the label of f with the labels of the vertices and edges surrounding f is equal to some fixed constant $\mu(s)$ for every s. We find necessary and sufficient conditions for (m, n, a, b, c) such that the calendula graph $Cal(m, n)$ admits a face-magic labeling of type (a, b, c).

Keywords Calendula graph · Face-magic graph labeling · H-magic graph labeling

MSC Codes 05C78

1 Introduction

Let $G = (V, E, F)$ be a planar graph and $a, b, c \in \{0, 1\}$. A *labeling of type* (a, b, c) is an assignment of the integers $1, 2, \ldots, a|V|+b|E|+c|F|$ to the elements of $V \cup E \cup F$ such that each vertex receives a labels, each edge receives b labels, and each face receives c labels, and no label is repeated. The *weight* of a face in a type (a, b, c) labeled graph is the sum of its own face label and the labels of its

B. Freyberg (✉) · G. Jensen · K. Scott
Department of Mathematics and Statistics, University of Minnesota Duluth, Duluth, MN, USA
e-mail: frey0031@d.umn.edu

A. Bahmanian (ed.), *Combinatorics and Combinatorial Computing*, Springer Proceedings in Mathematics & Statistics 516,
https://doi.org/10.1007/978-3-032-08549-8_7

for $i = 1, 2, \ldots, n-1$, and

$$\ell(v_n v_1) = \{\frac{n^2+2n}{2}, \frac{n^2-4n+2}{2}\}.$$

All of the labels from B have now been assigned. Form $\frac{n(n-4)}{2}$ complementary pairs p, q such that $p \in A$, $q \in C$, and $p + q = n^2 - n + 1$. Complete the labeling ℓ by arbitrarily assigning $\frac{n-4}{2}$ such pairs to each of the n edges of G.

We compute the weight of each edge next. Let $e = v_i v_{i+1} \in E(G)$ be given. If $i \in [1, n-1]$, we have

$$\begin{aligned} w(e) &= \ell(v_i) + \ell(v_{i+1}) + \ell(e) \\ &= (n^2 - 2n + 2i + 1) + (n^2 + 1 - 2i) + \tfrac{n-4}{2}(n^2 - n + 1) \\ &= \tfrac{n}{2}(n^2 - n + 1), \end{aligned}$$

and if $i = n$, we have

$$\begin{aligned} w(e) &= (n^2 - 2n + n + 1) + (n^2 - 2n + 1) + \tfrac{n-4}{2}(n^2 - n + 1) \\ &= \tfrac{n}{2}(n^2 - n + 1). \end{aligned}$$

Therefore, ℓ is an EMT labeling. It remains to show that the sum of the vertex labels is also $\frac{n}{2}(n^2 - n + 1)$. Indeed,

$$\sum_{i=1}^{n} \ell(v_i) = \frac{n(n^2-2n)}{2} + \frac{n(n+1)}{2} = \frac{n}{2}(n^2 - n + 1),$$

so the proof is complete. □

2.2 *Magic rectangles and quasi-magic rectangles*

A *magic square* is an $m \times m$ array of distinct integers $1, 2, \ldots, m^2$ in which the row, column, and diagonal sums are equal. It is well known that a magic square exists for every order m.

A generalization of the magic square, an $m \times n$ *magic rectangle* is an $m \times n$ array of distinct integers $1, 2, \ldots, mn$ so that the row sums are all equal and the column sums are all equal. Harmuth proved the following nearly 150 years ago [5],[6].

Theorem (Harmuth [5, 6]) *An $m \times n$ magic rectangle exists if and only if $m \equiv n$ (mod 2), except if $m = 1$, $n = 1$, or $m = n = 2$.* □

Therefore, when m and n are of different parities, an $m \times n$ magic rectangle does not exist. However, Froncek, Paramasivam, and Prajeesh addressed this problem by introducing quasi-magic rectangles [3].

Let m be an odd integer and n an even integer. An $m \times n$ *quasi-magic rectangle* $QMR(m, n : \lambda)$ is an m by n array such that each of the integers $1, 2, \ldots, \lambda-1, \lambda+1, \ldots, mn+1$ appears exactly once, the sum of the entries in each row is equal to some constant ρ, and the sum of the entries in each column is equal to some constant σ. Their result most pertinent to this work follows.

Theorem (Froncek et al. [3]) *A quasi-magic rectangle $QMR(m, 2t : mt+1)$ exists for all odd $m \geq 1$ and $t \geq 1$ except when $t = 1$ and $m \equiv 1 \pmod 4$.* □

2.3 *Constant sum partitions*

A fundamental problem many magic-type labelings have in common is the ability to partition a set into subsets with a constant sum. We call such a partition a *constant sum partition*. We are particularly concerned about finding constant sum partitions in which the subsets have equal size. Miller et al. showed the following in [9].

Theorem (Miller et al. [9]) *Let $m, n \geq 2$ be given and $S = [1, mn]$. There exists a constant sum partition of S into n subsets of size m if and only if m is even or both m and n are odd.* □

Notice the previous result means if n is even and m is odd, there does not exists a constant sum partition of $S = [1, mn]$ into n subsets of size m. However, if we include $mn+1$ in S and exclude some other member of S, the following corollary of Theorem 2.2 allows us to find such a constant sum partition.

Corollary 2.1 (Froncek et al. [3]) Let $n \geq 4$ be even, $m \geq 3$ be odd, and $S = \{1, 2, \ldots, \frac{mn}{2}, \frac{mn}{2}+2, \ldots, mn+1]$. There exists a constant sum partition of S into n subsets of size m. □

3 Main results

We will use the notation from Fig. 1 to denote the vertices, edges, and faces of $Cal(m, n)$. Notice that there are $m(n-1)$ vertices, mn edges, and $m+2$ faces. The m peripheral faces have n sides, the inner face F_0 has m sides, and the exterior face F_∞ has mn sides. Since $m, n \geq 3$, we can ignore the weight of the exterior face F_∞ since $mn \neq m$ and $mn \neq n$.

3.1 *$m \neq n$*

We begin by considering calendula graphs $Cal(m, n)$ with $m \neq n$. In this case, we need only concern ourselves with the m peripheral n-sided faces since the number of sides on the two remaining faces (the inner m-sided face and the exterior $m(n-1)$-sided face), both differ from n.

Lemma 3.1 Let $m, n \geq 3$, $m \neq n$, and $b, c \in \{0, 1\}$ be given. The calendula graph $Cal(m, n)$ admits a face-magic labeling of type $(0, b, c)$ except when

- $b = 0$ and $c = 1$ or
- $b = 1, c = 0$, m is even, and n is odd.

□

Proof Let $G \cong Cal(m, n)$ be embedded in the plane as in Fig. 1. It is obvious that no graph with 2 or more faces admits a face-magic labeling of type $(0, 0, 1)$ since the weight of a face would be equal to its label. So we may assume $(b, c) \neq (0, 1)$. This leaves two cases to consider.

Case 1 Type $(0, 1, 0)$ □

Notice that a face-magic labeling of this type corresponds to a partition of $[1, mn]$ into m constant sum sets of size n. Such a partition exists if and only if n is even or both m and n are odd by Theorem 2.3.

Case 2 Type $(0, 1, 1)$ □

Similar to the previous case, if n is odd or both m is odd and n is even, we may label the edges and peripheral faces using a constant sum partition of the set $[1, m(n+1)]$ into m subsets of size $n + 1$ which exists by Theorem 2.3. Then we may apply the labels $m(n + 1) + 1$ and $m(n + 1) + 2$ to the inner face and exterior face in any way we choose.

On the other hand, if both m and n are even, there exists a constant sum partition of $[1, m(n+1)+1] \setminus \{\frac{m(n+1)}{2}\}$ into m subsets of size $n+1$ by Corollary 2.1. We use this partition to label the edges and peripheral faces, and then we arbitrarily apply the labels $\frac{m(n+1)}{2}$ and $m(n + 1) + 2$ to the inner face and the exterior face. Since the weights of the inner face and exterior face are irrelevant, we are done. □

Next we prove the main result for this subsection.

Theorem *Let $m, n \geq 3$, $m \neq n$, and $a, b, c \in \{0, 1\}$ be given. The calendula graph $Cal(m, n)$ admits a type (a, b, c) face-magic labeling unless*

- $(a, b, c) = (0, 0, 1)$ *or*
- $(a, b, c) = (0, 1, 0)$*, m is even, and n is odd.*

□

Proof Let $G \cong Cal(m, n)$ be embedded in the plane as in Fig. 1. If $a = 0$, we are done by Lemma 3.1, so we may assume $a = 1$.

For the cases that follow let $H \cong C_m = (x_1, x_2, \ldots, x_m, x_1)$ and f be a type $(1, \beta)$ edge-magic total labeling of C_m with magic constant k_β; the specific value of β will be specified in each case and k is dependent upon β. Therefore, $f(x_i) \cup f(x_i x_{i+1}) \cup f(x_{i+1})$ is a set of $\beta+2$ integers that sum to k_β for every $i = 1, 2, \ldots, m$ (with arithmetic in the subscript taken modulo m). Furthermore, these sets partition

$$w(F_i) = \sum_{j=1}^{m+1} q_{i,j} = \rho.$$

The weight of the inner face is

$$w(F_0) = \sum_{i=1}^{m} q_{i,1} = \sigma + \lambda = \rho.$$

Since the labeling described is clearly a bijection between $E(G) \cup F(G)$ and $[1, m^2+m+2]$, and the weight of every m-sided face is the same, we have described a type (0, 1, 1) face-magic labeling of G.

Case 10 Type $(1, 0, 1)$ and m odd □

Let $S = [1, m^2 + 2]$ be the set of labels. If $m = 3$, label the inner triangle $\ell(v_1^1) = 6, \ell(v_1^2) = 7, \ell(v_1^3) = 8$, and $\ell(F_0) = 5$. This gives $w(F_0) = 26$. Then label the peripheral triangles, $\ell(v_2^1) = 2, \ell(F_1) = 11, \ell(v_2^2) = 1, \ell(F_2) = 10$, and $\ell(v_2^3) = 3, \ell(F_3) = 9$. Finally, let $\ell(F_\infty) = 4$. One can check that the weight of all three peripheral triangles is also 26, so ℓ is face-magic.

From now on assume $m \geq 5$ and label the vertices and faces

$$\ell(v_1^i) = \frac{m(m-1)}{2} + 2 + i,$$

and

$$\ell(F_i) = \frac{m(m+3)}{2} + 3 - i$$

for $i = 1, 2, \ldots, m$, and

$$\ell(v_2^i) = \frac{m(m+5)}{2} + 2 - i,$$

for $i = 1, 2, \ldots, m-1$. Then let $\ell(v_2^m) = \frac{m(m+5)}{2} + 2, \ell(F_0) = \frac{m(m-1)}{2}$ and $\ell(F_\infty) = m+1$. So far, we have assigned the labels $[\frac{m(m-1)}{2} + 2, \frac{m(m+5)}{2} + 2] \cup \{m+1\}$ to G and it only remains to assign labels to the remaining $m-3$ vertices on each peripheral face. We describe how to do that next.

Partition the unused labels of S into sets X, X', Y, and Y' where $X = [1, m]$, $X' = [m+2, 2m+1]$, $Y = [2m+2, \frac{m(m-1)}{2} + 1]$, and $Y' = [\frac{m(m+5)}{2} + 3, m^2 + 2.]$ Notice that $|X| = |X'| = m$ and $|Y| = |Y'| = \frac{m(m-5)}{2}$. Then form m pairs (x, x') such that $x \in X$ and $x' \in X'$ so that $x + x' = 2m + 2$. Similarly, form $\frac{m(m-5)}{2}$ pairs (y, y') such that $y \in Y$, $y' \in Y'$, and $y + y' = m^2 + 2m + 4$. Complete the labeling ℓ by arbitrarily assigning 1 (x, x') pair and $\frac{m-5}{2}$ (y, y') pairs to the remaining $m-3$

vertices of each peripheral face. Clearly $\ell : V \cup F \to S$ is a bijection, so it only remains to check the weights. The weight of the inner face is

$$\begin{aligned} w(F_0) &= \ell(F_0) + \textstyle\sum_{i=1}^{m} \ell(v_1^i) \\ &= \tfrac{m(m-1)}{2} + 2 + m(2 + \tfrac{m(m-1)}{2}) + \textstyle\sum_{i=1}^{m} i \\ &= \tfrac{(m+1)(m^2+4)}{2}. \end{aligned}$$

Let $i \in [1, m-1]$. The weight of the ith peripheral face is

$$\begin{aligned} w(F_i) &= \ell(v_1^i) + \ell(v_1^{i+1}) + \ell(v_2^i) + \ell(F_i) + (2m+2) + \tfrac{m-5}{2}(m^2+2m+4) \\ &= (\tfrac{m(m-1)}{2} + 2 + i) + (\tfrac{m(m-1)}{2} + 3 + i) + (\tfrac{m(m+5)}{2} + 2 - i) + \\ &\quad (\tfrac{m(m+3)}{2} + 3 - i) + (2m+2) + \tfrac{m-5}{2}(m^2+2m+4) \\ &= \tfrac{(m+1)(m^2+4)}{2}. \end{aligned}$$

Finally, the weight of the mth peripheral face is

$$\begin{aligned} w(F_m) &= \ell(v_1^m) + \ell(v_1^1) + \ell(v_2^m) + \ell(F_m) + (2m+2) + \tfrac{m-5}{2}(m^2+2m+4) \\ &= (\tfrac{m(m-1)}{2} + 2 + m) + (\tfrac{m(m-1)}{2} + 3) + (\tfrac{m(m+5)}{2} + 2) + \\ &\quad (\tfrac{m(m+3)}{2} + 3 - m) + (2m+2) + \tfrac{m-5}{2}(m^2+2m+4) \\ &= \tfrac{(m+1)(m^2+4)}{2}. \end{aligned}$$

Hence, ℓ is face-magic. □

Case 11 Type (1, 0, 1) and m even

Let $m = 2t$. If t is even, let

$$\ell(v_1^i) = \begin{cases} 2t^2 + 2t + 2 - \frac{i+1}{2} & \text{for } i = 1, 3, \ldots, t+1 \\ 2t^2 - t + 2 & \text{for } i = 2 \\ 2t^2 + t + 2 - \frac{i}{2} & \text{for } i = 4, 6, \ldots, t \\ 2t^2 + 2t + 2 - \frac{i+2}{2} & \text{for } i = t+2, t+4, \ldots, 2t \\ 2t^2 + t - \frac{i-5}{2} & \text{for } i = t+3, t+5, \ldots, 2t-1. \end{cases}$$

If t is odd, let

$$\ell(v_1^i) = \begin{cases} 2t^2 + 2t + 2 - \frac{i+1}{2} & \text{for } i = 1, 3, \ldots, t \\ 2t^2 - t + 2 & \text{for } i = 2 \\ 2t^2 + t + 1 - \frac{i}{2} & \text{for } i = 4, 6, \ldots, t-1 \\ \frac{4t^2+3t+1}{2} & \text{for } i = t+1 \\ 2t^2 + 2t + 2 - \frac{i+3}{2} & \text{for } i = t+2, t+4, \ldots, 2t-1 \\ 2t^2 + t + 2 - \frac{i}{2} & \text{for } i = t+3, t+5, \ldots, 2t-2 \\ 2t^2 + t & \text{for } i = 2t. \end{cases}$$

Let $\ell(F_0) = m + 1$ and $\ell(F_\infty) = m^2 - m + 2$. Then label each peripheral face

$$\ell(F_i) = \frac{3m^2 + m + 8}{2} - \ell(v_i^i) - \ell(v_1^{i+1})$$

for $i = 1, 2, \ldots, m$. The remaining labels constitute the sets

$$\begin{aligned} X &= [1, m] \\ Y &= [m + 2, \tfrac{m^2-2m+2}{2}] \\ Y' &= [\tfrac{m^2+2m+4}{2}, m^2 - m + 1] \\ X' &= [m^2 - m + 3, m^2 + 2] \end{aligned}$$

Notice that $|X| = |X'| = m$ and $|Y| = |Y'| = \frac{m^2-m}{2}$, so we may form m pairs (x, x') such that $x \in X$, $x' \in X'$, and $x + x' = m^2 + 3$. Similarly, we may form $\frac{m^2-4m}{2}$ pairs (y, y') with $y \in Y$, $y' \in Y'$, and $y + y' = m^2 + 3$. We complete the labeling ℓ by arbitrarily assigning these pairs to the unlabeled $m-2$ vertices on each of the m peripheral faces.

The weight of the inner face is

$$w(F_0) = \ell(F_0) + \sum_{i=1}^{m} \ell(v_1^i)$$

If t is even, this is

$$w(F_0) = (2t + 1) + (4t^3 + 3t^2 + 2t + 2) - (t^2 + 2)$$

while if t is odd, we have

$$w(F_0) = (2t + 1) + (4t^3 + 3t^2 + 2t - 2) - (t^2 - 2).$$

In either case,

$$\begin{aligned} w(F_0) &= 4t^3 + 2t^2 + 4t + 1 \\ &= \tfrac{m^3+m^2+4m+2}{2}. \end{aligned}$$

The weight of a peripheral face F_i is

$$\begin{aligned} w(F_i) &= \ell(F_i) + \ell(v_1^i) + \ell(v_1^{i+1}) + \tfrac{m-2}{2}(m^2 + 3) \\ &= \tfrac{3m^2+m+8}{2} + \tfrac{m-2}{2}(m^2 + 3) \\ &= \tfrac{m^3+m^2+4m+2}{2}. \end{aligned}$$

Hence, ℓ is face-magic.

Double-Change Covering Designs with Block Size $k = 2$ or 3

Nirosh T. Gamachchige, John P. McSorley, and Kristen A. Ceballos

Abstract A double-change covering design (DCCD(v, k)) is an ordered collection of b blocks, $\mathcal{B} = (B_1, B_2, \ldots, B_b)$, where each block B_s is a k-subset ($k \geq 2$) of $[v] = \{1, 2, \ldots, v\}$, such that: 1. every unordered pair of $[v]$ is covered by at least one block, and 2. each block differs from the previous block by two elements. The objective is to minimize b for a fixed v and k. Such a $\mathcal{B} = (B_1, B_2, \ldots, B_b)$ is called a linear DCCD(v, k). In a circular DCCD(v, k) block B_1 also differs from block B_b by two elements. For $k = 2$ and 3 we construct linear and circular DCCD(v, k)'s with the smallest possible number of blocks.

Keywords Double-change · Covering · Design · Blocks · Ordered

MSC Code: 05B30

1 Double-Change Covering Designs

1.1 Introduction, Definitions, Notation, an Example

Let $[v] = \{1, 2, \ldots, v\}$ and $\mathcal{B} = (B_1, B_2, \ldots, B_b)$ be an ordered collection of b blocks, where each block B_s is a k-subset ($k \geq 2$) of $[v]$, such that each block differs from the previous block by two elements, i.e., $|B_{s-1} \cap B_s| = k - 2$ for $s = 2, 3, \ldots, b$. We say there is a *double-change* between blocks in $\mathcal{B}$.

N. T. Gamachchige
Department of Science and Technology, Uva Wellassa University, Badulla, Sri Lanka
e-mail: tharaka@uwu.ac.lk

J. P. McSorley (✉) · K. A. Ceballos
Department of Mathematics, Southern Illinois University, Carbondale, IL, USA
e-mail: siu7684@siu.edu; kalindb@siu.edu

A. Bahmanian (ed.), *Combinatorics and Combinatorial Computing*, Springer Proceedings in Mathematics & Statistics 516,
https://doi.org/10.1007/978-3-032-08549-8_8

The first block B_1 covers all pairs in B_1. For $s = 2, 3, \ldots, b$, if $x, y \in B_s$ and $x, y \notin B_{s-1}$ then x, y are *introduced* in B_s, and the pairs covered by B_s are $\{x, y\} \cup \{x, z\} \cup \{y, z\}$ for $z \in B_s \setminus \{x, y\}$.

A *double-change covering design* (DCCD(v, k)) $\mathcal{D}$ based on $[v] = \{1, 2, \ldots, v\}$ with block size k is an ordered collection of b blocks, $\mathcal{B} = (B_1, B_2, \ldots, B_b)$, where block B_s is a k-subset of $[v]$, such that:

1. every unordered pair $\{x, y\}$ of $[v]$, with $x \neq y$, is covered by some block B_s with $s = 1, 2, \ldots, b$;
2. there is a double-change between blocks B_{s-1} and B_s for $s = 2, 3, \ldots, b$.

We say that pair $\{x, y\}$ is covered by $\mathcal{B}$ if it is covered by some block $B_s \in \mathcal{B}$.

Clearly $v \geq k$, and if $v = k$ we call $\mathcal{D}$ *trivial.* Also, if $k = v - 1$, then a double-change between blocks is not possible, so $2 \leq k \leq v - 2$, i.e., $v \geq k + 2$.

In a *linear* DCCD(v, k) all elements of B_1 are introduced in B_1, and two elements are introduced in B_s for each $s = 2, 3, \ldots, b$.

In a *circular* DCCD(v, k) there is also a double-change between blocks B_b and B_1, i.e., $|B_b \cap B_1| = k - 2$. Thus, in a circular DCCD(v, k), two elements are introduced per block, for every block.

We denote a linear DCCD(v, k) based on $[v]$ with block size k by *linear* $DCCD(v, k)$, or by *linear* $DCCD(v, k, b)$ if we wish to mention that it contains b blocks. Similarly for circular DCCD(v, k), we use *circular* $DCCD(v, k)$ or *circular* $DCCD(v, k, b)$.

Example 1 A linear DCCD(7, 3) or linear DCCD(7, 3, 7), $\mathcal{D}$, based on [1–7].

Blocks	$\mathcal{D}$	Elements introduced	Pairs covered
B_1	1 2 3	1,2,3	$\{1, 3\}, \{1, 2\}, \{2, 3\}$
B_2	1 4 6	4,6	$\{4, 6\}, \{1, 4\}, \{1, 6\}$
B_3	1 5 7	5,7	$\{5, 7\}, \{1, 5\}, \{1, 7\}$
B_4	3 5 4	3,4	$\{3, 4\}, \{3, 5\}, \{4, 5\}$
B_5	2 5 6	2,6	$\{2, 6\}, \{2, 5\}, \{5, 6\}$
B_6	3 7 6	3,7	$\{3, 7\}, \{3, 6\}, \{6, 7\}$
B_7	2 7 4	2,4	$\{2, 4\}, \{2, 7\}, \{4, 7\}$

Design $\mathcal{D}$ can also be seen as a circular DCCD(7, 3, 7), in which the elements introduced in B_1 are $1, 3$, and the pairs covered by B_1 are $\{1, 3\}$, $\{1, 2\}$, and $\{2, 3\}$.

Remark 1 A circular DCCD(v, k, b) can also be seen as a linear DCCD(v, k, b); so, in constructing such designs for a given v and k, we often attempt to first construct a circular DCCD(v, k, b).

1.2 History

The theory and construction of *single-change* covering designs have been widely discussed by mathematicians; see Wallis et al. [11]. Double-change covering designs were first mentioned in Preece et al. [10], but not considered further. Many of the results in this paper come from the MSc. Research Papers of Gamachchige [6] and Lindbloom [8]. Further recent results are discussed in the MSc. Thesis of Chafee [1]. See also Chafee and Stevens [2].

Dimitrov studied linear double-change covering designs and submitted a report [4]. His definition is different from ours in that it doesn't count the introduced pair $\{x, y\}$ in B_s as being covered in B_s: the pairs covered by B_s are $\{x, z\} \cup \{y, z\}$ for $z \in B_s \setminus \{x, y\}$.

Double-change covering designs can be applied in testing compatibility of electrical components. For example, suppose there are v different electrical components that need to be tested, each component must be tested with every other component. A number of components, say 1, 2, 3, are loaded into a testing device and tested concurrently; hence the components in the device form a block. At the completion of the test, two of the components are removed and replaced with two other components, thus forming the second block, and so on.

A linear or circular DCCD(v, k) is *tight* if every pair from $[v]$ is covered exactly once. In Theorem 2.1 of Horak and Rosa [7] it is proved that the block intersection graph of a Steiner 2- design S$(2, k, v)$ is Hamiltonian. This confirms the existence of a tight circular DCCD$(v, 3)$ for $v \equiv 1$ or $3 \pmod 6$.

1.3 Minimum Number of Blocks in a DCCD(v, k)

Consider a DCCD(v, k) with points $[v] = \{1, 2, \ldots, v\}$ and an ordered collection of b blocks $(B_1, B_2, \ldots, B_b)$.

We denote by $b^l_m(v, k)$ the minimum b for which there exists a linear DCCD(v, k, b), and $b^c_m(v, k)$ the minimum b for which there exists a circular DCCD(v, k, b). Our main objective is to determine $b^l_m(v, k)$ and $b^c_m(v, k)$, and to construct the corresponding minimum designs.

Theorem 1

(i) For a linear design we have

$$b^l_m(v, k) \geq \left\lceil \frac{\binom{v}{2} - \binom{k}{2}}{2k - 3} \right\rceil + 1.$$

(ii) For a circular design we have

$$b^c_m(v,k) \geq \left\lceil \frac{\binom{v}{2}}{2k-3} \right\rceil.$$

Proof

(i) A minimum linear DCCD$(v, k, b^l_m(v,k))$ must cover $\binom{v}{2}$ pairs. The first block covers $\binom{k}{2}$ pairs and each of the remaining $b^l_m(v,k)-1$ blocks covers $1+2(k-2) = 2k-3$ pairs. Thus $(b^l_m(v,k)-1)(2k-3)+\binom{k}{2} \geq \binom{v}{2}$, and the result follows.

(ii) Similarly, a minimum circular DCCD$(v, k, b^c_m(v,k))$ must cover $\binom{v}{2}$ pairs. Here $2k-3$ pairs are covered per block. Thus $b^c_m(v,k)(2k-3) \geq \binom{v}{2}$, and the result follows.

□

We define the following two functions $b^l_e(v,k)$ and $b^c_e(v,k)$ using Theorem 1(i) and (ii):

Definition 1

$$(i)\ b^l_e(v,k) = \left\lceil \frac{\binom{v}{2}-\binom{k}{2}}{2k-3} \right\rceil + 1 \quad \text{and} \quad (ii)\ b^c_e(v,k) = \left\lceil \frac{\binom{v}{2}}{2k-3} \right\rceil. \tag{1}$$

For a linear design if $b^l_m(v,k) = b^l_e(v,k)$, or for a circular design if $b^c_m(v,k) = b^c_e(v,k)$, then the corresponding design is called *economical.*

A (i) linear or (ii) circular design is *tight* if every pair from $[v]$ is covered exactly once, so

$$(i)\ b^l_m(v,k) = \frac{\binom{v}{2}-\binom{k}{2}}{2k-3} + 1 \quad \text{or} \quad (ii)\ b^c_m(v,k) = \frac{\binom{v}{2}}{2k-3}. \tag{2}$$

If $(2k-3) \mid (\binom{v}{2}-\binom{k}{2})$ for the linear case or $(2k-3) \mid \binom{v}{2}$ for the circular case, we call (v,k) a *tight case.* Clearly any tight DCCD(v,k) is an economical DCCD(v,k), and any economical DCCD(v,k) is a minimum DCCD(v,k).

A DCCD(v,k) $\mathcal{D}$ is *regular* if each element $x \in [v]$ is contained in the same number of blocks of $\mathcal{D}$.

Here is a summary of this paper:

In Sect. 2 we constuct a tight circular DCCD$(v, 2)$ for all $v \geq 6$.

In Sect. 3 we construct minimum DCCD$(v, 3)$'s for each of the 6 modulo classes of $v \pmod 6$. Namely; tight DCCD$(v, 3)$'s when $v \equiv 1$ or $3 \pmod 6$, economical DCCD$(v, 3)$'s for $v \equiv 5 \pmod 6$, and minimum DCCD$(v, 3)$'s for $v \equiv 0, 2,$ or $4 \pmod 6$.

In Sect. 4 we consider regular DCCD(v,k)'s for $k = 2$ and 3.

In future papers we will consider both linear and circular DCCD(v,k)'s for general block size k, with emphasis on $k = 4$.

2 Double-Change Covering Designs with $k = 2$

From Eqs. (1) and (2) with $k = 2$ we have $b_e^l(v, 2) = b_e^c(v, 2) = \binom{v}{2}$, and each $(v, 2)$ is a tight case. In accordance with Remark 1 we will consider the existence/non-existence of a tight circular DCCD$(v, 2, \binom{v}{2})$.

We start with $v = k + 2 = 4$, it is straightforward to see that there is no linear nor circular DCCD$(4, 2)$.

For $v = 5$ there is no tight circular DCCD$(5, 2, 10)$. Suppose that a tight circular DCCD$(5, 2, 10)$ exists. Without loss of generality, let $B_1 = \{1, 2\}$, $B_{10} = \{3, 4\}$, $B_2 = \{3, 5\}$, and $B_9 = \{2, 5\}$ (see Fig. 1a). Then we can easily check that the remaining blocks (pairs) cannot be fitted in. However there is a minimum circular DCCD$(5, 2, 11)$ shown in Fig. 1b, and a tight linear DCCD$(5, 2, 10)$ is obtained by removing the last block of (b).

Definition 2 Let G be a simple graph, denote by $E(G)$ the set of edges of G. Two edges are *independent* if they have no common end vertex. A set of d independent edges of G is called a *d-matching* of G.

Theorem 2 *For $v \geq 6$ a tight circular and tight linear DCCD$(v, 2, \binom{v}{2})$ exists.*

Proof Consider the complete graph K_v on vertices $[v]$. We identify blocks of a DCCD$(v, 2)$ with edges of K_v. To form a tight DCCD$(v, 2, \binom{v}{2})$ we partition $E(K_v)$ into suitable matchings, and then arrange the edges in the matchings so no two successive edges have a vertex in common.

First, if $v \geq 6$ is even, we partition $E(K_v)$ into $v - 1$ many $\frac{v}{2}$ - matchings as follows:

$$E(K_v) = \bigcup_{i=0}^{v-2} \{\{\sigma^i(1), \sigma^i(v-1)\}, \{\sigma^i(2), \sigma^i(v-2)\}, \ldots, \{\sigma^i\left(\frac{v}{2}\right), \sigma^i(v)\}\}$$

Fig. 1 (**a**) An arrangement of pairs for a possible tight circular DCCD(5, 2, 10). (**b**) A minimum circular DCCD(5, 2, 11). (**c**) A tight circular DCCD(6, 2, 15), see Example 2

		1 5
		2 4
		3 6
		2 1
	1 2	3 5
1 2	3 4	4 6
3 5	1 5	3 2
B_3	2 3	4 1
B_4	4 5	5 6
B_5	1 3	4 3
B_6	2 5	5 2
B_7	1 4	1 6
B_8	3 5	5 4
2 5	2 4	1 3
3 4	3 5	2 6
(a)	(b)	(c)

where $\sigma = (12 \cdots v-1)(v)$.

If $v \geq 7$ is odd, we partition $E(K_v)$ into v $\frac{v-1}{2}$-matchings:

$$E(K_v) = \bigcup_{i=0}^{v-1} \{\{\sigma^i(1), \sigma^i(v-1)\}, \{\sigma^i(2), \sigma^i(v-2)\}, \ldots, \{\sigma^i\left(\frac{v-1}{2}\right), \sigma^i\left(\frac{v+1}{2}\right)\}\}$$

where $\sigma = (12 \cdots v)$.

In both cases we choose successive blocks by moving along the rows of the matchings, see Example 2 below. It is straightforward to show that we obtain a tight circular DCCD(v, 2, $\binom{v}{2}$) which is also a tight linear DCCD(v, 2, $\binom{v}{2}$). □

Example 2 For $v = 6$, the following table gives the permutations σ^i and corresponding 3-matchings.

Permutation	3-matching
(1)(2)(3)(4)(5)(6)	{1,5}, {2,4}, {3,6}
(1 2 3 4 5)(6)	{2,1}, {3,5}, {4,6}
(1 3 5 2 4)(6)	{3,2}, {4,1}, {5,6}
(1 4 2 5 3)(6)	{4,3}, {5,2}, {1,6}
(1 5 4 3 2)(6)	{5,4}, {1,3}, {2,6}

Then Fig. 1c is the tight circular DCCD(6, 2, 15) obtained by arranging the edges of the 3-matchings.

3 Double-Change Covering Designs with $k = 3$

In light of Remark 1, in this section we construct minimum circular DCCD(v, 3)'s for each of the 6 modulo classes of $v \pmod 6$.

See Section 2 of Fort and Hedlund [5] and Section VI.11.3, Theorem 11.19 of Colbourn and Dinitz [3], the Schönheim Bound:

Theorem 3 *The minimum number of blocks, $b_m^c(v, 3)$, in a circular DCCD(v, 3) satisfies:*

$$b_m^c(v, 3) \geq \left\lceil \frac{v}{3} \left\lceil \frac{v-1}{2} \right\rceil \right\rceil.$$

Proof Let $\mathcal{D}$ be a minimum circular DCCD(v, 3, $b_m(v, 3)$). For each occurrence of x in $\mathcal{D}$ exactly 2 pairs containing x are covered by $\mathcal{D}$. Let x occur in block $B = \{x, y, z\}$. No matter which two elements in B are introduced in B, it is $\{x, y\}$ and $\{x, z\}$ that are the pairs covered by B that contain x. Now there are $|V \setminus \{x\}| = v - 1$

pairs containing x to be covered by $\mathcal{D}$, hence each x occurs at least $\left\lceil \frac{v-1}{2} \right\rceil$ times in $\mathcal{D}$. There are v such elements x, and each block has size 3, this gives the result. □

Let λK_v, for some $\lambda \geq 1$, be the multi-graph on v vertices in which each pair of vertices is joined by λ edges.

See Section 4.1 of Lindner and Rodger [9] for the following definition, except that the word minimum applies to $|\mathcal{T}|$ rather than to $|\mathcal{P}|$.

Definition 3 A *minimum covering with triangles* (MCT) of the complete graph K_v is a triple $(S, \mathcal{T}, \mathcal{P})$, where S is the vertex set of K_v, and $\mathcal{P}$ is a subset of $E(\lambda K_v)$, for some $\lambda \geq 1$, and $\mathcal{T}$ is a collection of triangles which partition $E(K_v) \cup \mathcal{P}$, and is as small as possible. The set of edges $\mathcal{P}$ is called the *padding*.

Remark 2 We identify a MCT of K_v with a minimum covering with triangles of the set $[v] = \{1, 2, \ldots, v\}$ by identifying vertices of K_v with elements of $[v]$ in the natural way, and denote such a covering by MCT(v).

We define

$$b_{\text{MCT}}(v, 3) = \left\lceil \frac{v}{3} \left\lceil \frac{v-1}{2} \right\rceil \right\rceil. \tag{3}$$

We will show that for each of the 6 modulo classes of $v \pmod 6$, the number of triangles (triples/blocks) in a MCT(v) is $b_{\text{MCT}}(v, 3)$, and that the blocks in $\mathcal{T}$ can be arranged in a circular double-change manner, thus giving a minimum circular DCCD$(v, 3, \left\lceil \frac{v}{3} \left\lceil \frac{v-1}{2} \right\rceil \right\rceil)$, which is also a minimum linear DCCD$(v, 3, \left\lceil \frac{v}{3} \left\lceil \frac{v-1}{2} \right\rceil \right\rceil)$ by Remark 1.

Each MCT(v) has a corresponding padding $\mathcal{P}$, which are given in Section 4.1, Figure 4.4 of [9] and Section VI.11.5, Table 11.41 of Colbourn and Dinitz [3].

3.1 A Tight Circular DCCD$(6n + 1, 3, 6n^2 + n)$

Let $v = 6n + 1$ then $b_{\text{MCT}}(6n + 1, 3) = b_e^c(6n + 1, 3) = 6n^2 + n$.

For $n \geq 1$ we take the Skolem construction of a STS$(6n + 1)$ from Section 1.3 of [9], and arrange its blocks to give a tight circular DCCD$(6n + 1, 3, 6n^2 + n)$. Its padding $\mathcal{P}$ is the empty set.

The Skolem Construction Let $v = 6n + 1$ and $Q = \{1, 2, \ldots, 2n\}$, and $(Q, \circ)$ be a half-idempotent commutative quasigroup of order $2n$ based on Q. Let $S = \{\infty\} \cup (Q \times \{1, 2, 3\})$, and define $\mathcal{T}$ to contain the following three Types of blocks:

Type 1: $\{(i, 1), (i, 2), (i, 3)\}$ for $1 \leq i \leq n$,
Type 2: $\{\infty, (n + i, 1), (i, 2)\}, \{\infty, (n + i, 2), (i, 3)\}, \{\infty, (n + i, 3), (i, 1)\}$ for $1 \leq i \leq n$,
Type 3: $\{(i, 1), (j, 1), (i \circ j, 2)\}, \{(i, 2), (j, 2), (i \circ j, 3)\}, \{(i, 3), (j, 3), (i \circ j, 1)\}$ for $1 \leq i < j \leq 2n$.

Then $(S, \mathcal{T})$ is a STS$(6n + 1)$.

We partition the blocks of Type 3. Let

- $\mathcal{B}_1$ be the blocks of Type 3 of the form $\{(i, 1), (j, 1), (i \circ j, 2)\}$ for $1 \leq i < j \leq 2n$,
- $\mathcal{B}_2$ be the blocks of Type 3 of the form $\{(i, 2), (j, 2), (i \circ j, 3)\}$ for $1 \leq i < j \leq 2n$,
- $\mathcal{B}_3$ be the blocks of Type 3 of the form $\{(i, 3), (j, 3), (i \circ j, 1)\}$ for $1 \leq i < j \leq 2n$.

Now arrange them using pair i and j, i.e., $\boxed{i, j}$, as follows:

Arrangement of B_t for $t = 1, 2,$ or 3 For any block in $\mathcal{B}_t$ for $t = 1, 2,$ or 3 the third element of each block is uniquely determined by its first two elements i and j. If we arrange the i and j, i.e., $\boxed{i, j}$, of the first two elements as a single-change design, as shown in Fig. 2, we will get blocks with a double-change, see Step 3 below.

The following Steps 1–4 construct a tight circular DCCD$(6n+1, 3, 6n^2+n)$ for every $n \geq 1$. (Appropriately modified interpretations of some Steps are needed for $n = 1$.)

- **Step 1:** Let $Q = \{1, 2, \ldots, 2n\}$. We use the following half-idempotent commutative quasigroup of order $2n$, $(Q, \circ)$, whose rows are formed from the back-circulant of $(1, n+1, 2, n+2, \ldots, i, n+i, \ldots, n, 2n)$ to form the blocks of Type 3.

$\circ$	1	2	3	4	$\cdots$	$2n-1$	$2n$
1	1	$n+1$	2	$n+2$	$\cdots$	n	$2n$
2	$n+1$	2	$n+2$	3	$\cdots$	$2n$	1
3	2	$n+2$	3	$n+3$	$\cdots$	1	$n+1$
4	$n+2$	3	$n+3$	4	$\cdots$	$n+1$	2
.	.	.	.	.	$\cdots$	.	.
.	.	.	.	.	$\cdots$	.	.
.	.	.	.	.	$\cdots$	.	.
$2n-1$	n	$2n$	1	$n+1$	$\cdots$	$n-1$	$2n-1$
$2n$	$2n$	1	$n+1$	2	$\cdots$	$2n-1$	n

- **Step 2:** Going around the cycle in Fig. 2, arrange the blocks in $\mathcal{B}_t$ for $t = 1, 2,$ and 3 as follows: For $\mathcal{B}_1$, start with pair $\boxed{1, 3}$, follow the arrows, and finish with $\boxed{1, 2}$, i.e., start with block $\{(1, 1), (3, 1), (1 \circ 3, 2)\} = \{(1, 1), (3, 1), (2, 2)\}$ and finish with block $\{(1, 1), (2, 1), (1 \circ 2, 2)\} = \{(1, 1), (2, 1), (n+1, 2)\}$. For both $\mathcal{B}_2$ and $\mathcal{B}_3$, start with pair $\boxed{1, 2}$ and finish with $\boxed{2, 3}$. See Fig. 3.
- **Step 3:** Insert the Type 1 blocks into the Type 2 blocks as follows: for each $i = 1, 2, \ldots, n$, place the Type 1 block $\{(i, 1), (i, 2), (i, 3)\}$, shown in **bold**,

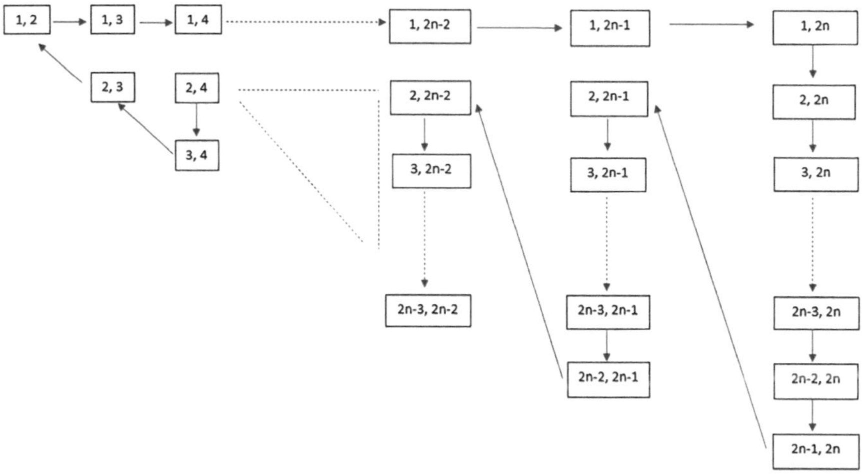

Fig. 2 Arrangement of $\boxed{i, j}$ for $\mathcal{B}_t$ for $t = 1, 2$, or 3

$(1,1), (3,1), \quad (2,2)$	$(1,2), (2,2), (\boldsymbol{n}+1,3)$	$(1,3), (2,3), (\boldsymbol{n}+1,1)$
$(1,1), (4,1), (\boldsymbol{n}+2,2)$	$(1,2), (3,2), \quad (2,3)$	$(1,3), (3,3), \quad (2,1)$
$\cdot \quad \cdot \quad \cdot$	$\cdot \quad \cdot \quad \cdot$	$\cdot \quad \cdot \quad \cdot$
$\cdot \quad \cdot \quad \cdot$	$\cdot \quad \cdot \quad \cdot$	$\cdot \quad \cdot \quad \cdot$
$\cdot \quad \cdot \quad \cdot$	$\cdot \quad \cdot \quad \cdot$	$\cdot \quad \cdot \quad \cdot$
$(1,1), (2,1), (\boldsymbol{n}+1,2)$	$(2,2), (3,2), (\boldsymbol{n}+2,3)$	$(2,3), (3,3), (\boldsymbol{n}+2,1)$
$\mathcal{B}_1$	$\mathcal{B}_2$	$\mathcal{B}_3$

Fig. 3 Blocks $\mathcal{B}_t$ for $t = 1, 2$, and 3

between the two Type 2 blocks $\{\infty, (n+i, 1), (i, 2)\}$ and $\{\infty, (n+i, 2), (i, 3)\}$. See Fig. 4a, the horizontal lines are included for clarity.

- **Step 4:** From Fig. 4a, insert the following blocks shown in **bold**:
  Insert the blocks $\mathcal{B}_1$ between B_2 and B_3, see the first $\triangleleft$ in Fig. 4a,
  insert the blocks $\mathcal{B}_2$ between B_5 and B_6, see the second $\triangleleft$ in Fig. 4a.
  insert the blocks $\mathcal{B}_3$ between B_6 and B_7, see the third $\triangleleft$ in Fig. 4a.
  See Fig. 4b.

We check that Fig. 4b is a tight circular DCCD$(6n + 1, 3, 6n^2 + n)$.

In this construction we have n blocks of Type 1, $3n$ blocks of Type 2, and $3n(2n-1)$ blocks of Type 3. Thus the number of blocks is $b = 6n^2 + n = b_{\text{MCT}}(6n + 1, 3)$, as needed.

See Theorem 2.1 of [7] and Theorem 3.2 of [1] for alternative proofs of existence of a tight circular DCCD$(6n + 1, 3, 6n^2 + n)$.

$$
\begin{array}{rl}
B_1 = & \infty, (n+1,1), (1,2) \\
B_2 = & \mathbf{(1,1), (1,2), (1,3)} \\
B_3 = & \infty, (n+1,2), (1,3)^{\triangleleft} \\
B_4 = & \infty, (n+1,3), (1,1) \\
\hline
B_5 = & \infty, (n+2,1), (2,2) \\
B_6 = & \mathbf{(2,1), (2,2), (2,3)}^{\triangleleft} \\
B_7 = & \infty, (n+2,2), (2,3)^{\triangleleft} \\
 & \infty, (n+2,3), (2,1) \\
\hline
 & \cdot \quad \cdot \quad \cdot \\
 & \cdot \quad \cdot \quad \cdot \\
 & \cdot \quad \cdot \quad \cdot \\
\hline
 & \infty, (n+i,1), (i,2) \\
 & \mathbf{(i,1), (i,2), (i,3)} \\
 & \infty, (n+i,2), (i,3) \\
 & \infty, (n+i,3), (i,1) \\
\hline
 & \cdot \quad \cdot \quad \cdot \\
 & \cdot \quad \cdot \quad \cdot \\
 & \cdot \quad \cdot \quad \cdot \\
\hline
 & \infty, (2n,1), (n,2) \\
 & \mathbf{(n,1), (n,2), (n,3)} \\
 & \infty, (2n,2), (n,3) \\
 & \infty, (2n,3), (n,1)
\end{array}
$$

(a)

$$
\begin{array}{rl}
B_1 = & \infty, (n+1,1), (1,2) \\
B_2 = & (1,1), (1,2), (1,3) \\
 & \mathbf{(1,1), (3,1), (2,2)} \\
 & \mathbf{(1,1), (4,1), (n+2,2)} \\
 & \cdot \quad \cdot \quad \cdot \\
 & \cdot \quad \cdot \quad \cdot \\
 & \cdot \quad \cdot \quad \cdot \\
 & \mathbf{(1,1), (2,1), (n+1,2)} \\
B_3 = & \infty, (n+1,2), (1,3) \\
B_4 = & \infty, (n+1,3), (1,1) \\
\hline
B_5 = & \infty, (n+2,1), (2,2) \\
 & \mathbf{(1,2), (2,2), (n+1,3)} \\
 & \mathbf{(1,2), (3,2), (2,3)} \\
 & \cdot \quad \cdot \quad \cdot \\
 & \cdot \quad \cdot \quad \cdot \\
 & \cdot \quad \cdot \quad \cdot \\
 & \mathbf{(2,2), (3,2), (n+2,3)} \\
B_6 = & (2,1), (2,2), (2,3)_{\triangleleft} \\
 & \mathbf{(1,3), (2,3), (n+1,1)} \\
 & \mathbf{(1,3), (3,3), (2,1)} \\
 & \cdot \quad \cdot \quad \cdot \\
 & \cdot \quad \cdot \quad \cdot \\
 & \cdot \quad \cdot \quad \cdot \\
 & \mathbf{(2,3), (3,3), (n+2,1)} \\
B_7 = & \infty, (n+2,2), (2,3) \\
 & \infty, (n+2,3), (2,1) \\
\hline
 & \cdot \quad \cdot \quad \cdot \\
 & \cdot \quad \cdot \quad \cdot \\
 & \cdot \quad \cdot \quad \cdot \\
\hline
 & \infty, (n+i,1), (i,2) \\
 & (i,1), (i,2), (i,3) \\
 & \infty, (n+i,2), (i,3) \\
 & \infty, (n+i,3), (i,1) \\
\hline
 & \cdot \quad \cdot \quad \cdot \\
 & \cdot \quad \cdot \quad \cdot \\
 & \cdot \quad \cdot \quad \cdot \\
\hline
 & \infty, (2n,1), (n,2) \\
 & (n,1), (n,2), (n,3) \\
 & \infty, (2n,2), (n,3) \\
 & \infty, (2n,3), (n,1)
\end{array}
$$

(b)

Fig. 4 (**a**) Block arrangement for Step 3. (**b**) Block arrangement for Step 4. A tight circular DCCD$(6n+1, 3, 6n^2+n)$

Example 3 $n = 2, v = 13, Q = \{1, 2, 3, 4\}$. We use the following half-idempotent commutative quasigroup of order 4, $(Q, \circ)$, and construct a tight circular DCCD(13, 3, 26).

$\circ$	1 2 3 4
1	1 3 2 4
2	3 2 4 1
3	2 4 1 3
4	4 1 3 2

Let $S = \{\infty\} \cup (\{1, 2, 3, 4\} \times \{1, 2, 3\}) = \{\infty, (1, 1), (1, 2), (1, 3), (2, 1), (2, 2), (2, 3), (3, 1), (3, 2), (3, 3), (4, 1), (4, 2), (4, 3)\}$. Then $\mathcal{T}$ contains the following 26 blocks, forming a STS(13):

Type 1: $i = 1, 2$ $\{(1, 1), (1, 2), (1, 3)\}, \{(2, 1), (2, 2), (2, 3)\}$,
Type 2: $i = 1$ $\{\infty, (3, 1), (1, 2)\}, \{\infty, (3, 2), (1, 3)\}, \{\infty, (3, 3), (1, 1)\}$,
$i = 1$ $\{\infty, (4, 1), (2, 2)\}, \{\infty, (4, 2), (2, 3)\}, \{\infty, (4, 3), (2, 1)\}$,
Type 3: $\mathcal{B}_1$ $\{(1, 1), (3, 1), (2, 2)\}, \{(1, 1), (4, 1), (4, 2)\}, \{(2, 1), (4, 1), (1, 2)\}$, $\{(3, 1), (4, 1), (3, 2)\}, \{(2, 1), (3, 1), (4, 2)\}, \{(1, 1), (2, 1), (3, 2)\}$,
$\mathcal{B}_2$ $\{(1, 2), (2, 2), (3, 3)\}, \{(1, 2), (3, 2), (2, 3)\}, \{(1, 2), (4, 2), (4, 3)\}$, $\{(2, 2), (4, 2), (1, 3)\}, \{(3, 2), (4, 2), (3, 3)\}, \{(2, 2), (3, 2), (4, 3)\}$,
$\mathcal{B}_3$ $\{(1, 3), (2, 3), (3, 1)\}, \{(1, 3), (3, 3), (2, 1)\}, \{(1, 3), (4, 3), (4, 1)\}$, $\{(2, 3), (4, 3), (1, 1)\}, \{(3, 3), (4, 3), (3, 1)\}, \{(2, 3), (3, 3), (4, 1)\}$.

Following the Steps 1–4 we obtain the tight circular DCCD(13, 3, 26):

∞, (3, 1), (1, 2)	∞, (4, 1), (2, 2)	(1, 3), (2, 3), (3, 1)
(1, 1), (1, 2), (1, 3)	(1, 2), (2, 2), (3, 3)	(1, 3), (3, 3), (2, 1)
(1, 1), (3, 1), (2, 2)	(1, 2), (3, 2), (2, 3)	(1, 3), (4, 3), (4, 1)
(1, 1), (4, 1), (4, 2)	(1, 2), (4, 2), (4, 3)	(2, 3), (4, 3), (1, 1)
(2, 1), (4, 1), (1, 2)	(2, 2), (4, 2), (1, 3)	(3, 3), (4, 3), (3, 1)
(3, 1), (4, 1), (3, 2)	(3, 2), (4, 2), (3, 3)	(2, 3), (3, 3), (4, 1)
(2, 1), (3, 1), (4, 2)	(2, 2), (3, 2), (4, 3)	∞, (4, 2), (2, 3)
(1, 1), (2, 1), (3, 2)	(2, 1), (2, 2), (2, 3) $\triangleleft$	∞, (4, 3), (2, 1)
∞, (3, 2), (1, 3)		
∞, (3, 3), (1, 1)		

3.2 A Minimum Circular DCCD($6n + 2, 3, 6n^2 + 4n + 1$)

Let $v = 6n + 2$ then $b_{\mathrm{MCT}}(6n + 2, 3) = 6n^2 + 4n + 1$.

We construct a minimum circular DCCD($6n+2, 3, 6n^2+4n+1$) by adding on a new point ∞', and appropriate blocks, to our tight circular DCCD($6n+1, 3, 6n^2+n$) of Fig. 4b (Sect. 3.1). Its padding is the tripole $\mathcal{P} = K_{1,3} \cup (3n - 1)K_2$.

For $n \geq 1$, with appropriate modifications of some parts for $n = 1$, see Section 4.3 of [9]:

The Tripole Covering Construction Let $v \equiv 2 \pmod 6$, and $(X, \mathcal{T})$ be a STS of order $v - 1 \equiv 1 \pmod 6$. Let $\{a, b, c\} \in \mathcal{T}$ be any block in $\mathcal{T}$, and let $\phi = \{\{a, b\}, \{b, c\}\}$, and let $\pi = \{\{x_1, y_1\}, \{x_2, y_2\}, \ldots, \{x_{\frac{v-4}{2}}, y_{\frac{v-4}{2}}\}\}$ be any partition of $X \setminus \{a, b, c\}$. Now let $\phi(\infty') = \{\{\infty', a, b\}, \{\infty', b, c\}\}$, and $\pi(\infty') = \{\{\infty', x_1, y_1\}, \{\infty', x_2, y_2\}, \ldots, \{\infty', x_{\frac{v-4}{2}}, y_{\frac{v-4}{2}}\}\}$. Further, let $S = \{\infty'\} \cup X$, and $\mathcal{T}^* = \mathcal{T} \cup \phi(\infty') \cup \pi(\infty')$, and $\mathcal{P} = \phi \cup \pi \cup \{\{\infty', b\}\}$.

Then $(S, \mathcal{T}^*, \mathcal{P})$ is a MCT(v).

Let $(X, \mathcal{T})$ be the tight circular DCCD$(6n + 1, 3, 6n^2 + n)$ of Fig. 4b, with $X = \{\infty\} \cup (Q \times \{1, 2, 3\})$. Let $B_2 = \{(1, 1), (1, 2), (1, 3)\} \in \mathcal{T}$ then $\phi = \{\{(1, 1), (1, 2)\}, \{(1, 2), (1, 3)\}\}$.

Let $\pi = \{\{(2i,1), (2i+1,1)\}, \{(2i,2), (2i+1,2)\}, \{(2i,3), (2i+1,3)\} (1 \le i \le n-1) \cup (\{(2n, 1), (2n, 2)\}, \{\infty, (2n, 3)\}\}$ be a partition of $X \setminus \{(1, 1), (1, 2), (1, 3)\}$.

- **Steps 1–4** of Sect. 3.1.
- **Step 5:** Arrange the blocks of $\pi(\infty')$ in a double-change manner. See Fig. 5a.
- **Step 6:** From Fig. 5a, insert the 2 blocks in $\phi(\infty')$, shown in **bold**, preserving the double-change. See Fig. 5b.
- **Step 7:** Insert the arrangement of blocks from Fig. 5b at the $\triangleleft$ (after B_6) in the tight circular DCCD$(6n + 1, 3, 6n^2 + n)$ of Fig. 4b to give a minimum circular DCCD$(6n + 2, 3, 6n^2 + 4n + 1)$.

Now $\phi(\infty')$ contains 2 blocks, and π is a partition of $X \setminus \{(1, 1), (1, 2), (1, 3)\}$, a $(6n - 2)$-set, into $3n - 1$ pairs, so $\pi(\infty')$ contains $3n - 1$ blocks. We check:

$$\begin{array}{lll}
\infty', & (2,1), & (3,1) \\
\infty', & (2,2), & (3,2) \\
\infty', & (2,3), & (3,3) \\
\cdot & \cdot & \cdot \\
\cdot & \cdot & \cdot \\
\cdot & \cdot & \cdot \\
\infty', & (2i,1), & (2i+1,1) \\
\infty', & (2i,2), & (2i+1,2) \\
\infty', & (2i,3), & (2i+1,3) \\
\cdot & \cdot & \cdot \\
\cdot & \cdot & \cdot \\
\cdot & \cdot & \cdot \\
\infty', & (2n-2,1), & (2n-1,1) \\
\infty', & (2n-2,2), & (2n-1,2) \\
\infty', & (2n-2,3), & (2n-1,3) \\
\infty', & (2n,1), & (2n,2) \\
\infty', & \infty, & (2n,3)
\end{array}
\qquad
\begin{array}{lll}
\infty', & (2,1), & (3,1) \\
\infty', & (2,2), & (3,2) \\
\infty', & (2,3), & (3,3) \\
\cdot & \cdot & \cdot \\
\cdot & \cdot & \cdot \\
\cdot & \cdot & \cdot \\
\infty', & (2i,1), & (2i+1,1) \\
\infty', & (2i,2), & (2i+1,2) \\
\infty', & (2i,3), & (2i+1,3) \\
\cdot & \cdot & \cdot \\
\cdot & \cdot & \cdot \\
\cdot & \cdot & \cdot \\
\infty', & (2n-2,1), & (2n-1,1) \\
\infty', & (2n-2,2), & (2n-1,2) \\
\infty', & (2n-2,3), & (2n-1,3) \\
\infty', & (2n,1), & (2n,2) \\
\boldsymbol{\infty'}, & \mathbf{(1,1)}, & \mathbf{(1,2)} \\
\infty', & \infty, & (2n,3) \\
\boldsymbol{\infty'}, & \mathbf{(1,2)}, & \mathbf{(1,3)}
\end{array}$$

(a) (b)

Fig. 5 (**a**) Block arrangement for Step 5. (**b**) Block arrangement for Step 6

$b_{\mathrm{MCT}}(6n+1, 3)+2+3n-1 = 6n^2+n+3n+1 = 6n^2+4n+1 = b_{\mathrm{MCT}}(6n+2, 3)$, as needed.

Example 4 $n = 2, v = 14$. We take the tight circular DCCD(13, 3, 26) of Example 3 and add on one extra point ∞', and the $3n + 1 = 7$ new blocks $\phi(\infty') \cup \pi(\infty')$:

$$\begin{array}{l} \infty', (1, 1), (1, 2) \\ \infty', (1, 2), (1, 3) \\ \infty', (2, 1), (3, 1) \\ \infty', (2, 2), (3, 2) \\ \infty', (2, 3), (3, 3) \\ \infty', (4, 1), (4, 2) \\ \infty', \ \infty, \ (4, 3) \end{array}$$

at the $\triangleleft$, to obtain the minimum circular DCCD(14, 3, 33):

∞, (3, 1), (1, 2)	∞, (4, 1), (2, 2)	(1, 3), (2, 3), (3, 1)
(1, 1), (1, 2), (1, 3)	(1, 2), (2, 2), (3, 3)	(1, 3), (3, 3), (2, 1)
(1, 1), (3, 1), (2, 2)	(1, 2), (3, 2), (2, 3)	(1, 3), (4, 3), (4, 1)
(1, 1), (4, 1), (4, 2)	(1, 2), (4, 2), (4, 3)	(2, 3), (4, 3), (1, 1)
(2, 1), (4, 1), (1, 2)	(2, 2), (4, 2), (1, 3)	(3, 3), (4, 3), (3, 1)
(3, 1), (4, 1), (3, 2)	(3, 2), (4, 2), (3, 3)	(2, 3), (3, 3), (4, 1)
(2, 1), (3, 1), (4, 2)	(2, 2), (3, 2), (4, 3)	∞, (4, 2), (2, 3)
(1, 1), (2, 1), (3, 2)	(2, 1), (2, 2), (2, 3)	∞, (4, 3), (2, 1)
∞, (3, 2), (1, 3)	**∞', (2, 1), (3, 1)**	
∞, (3, 3), (1, 1)	**∞', (2, 2), (3, 2)**	
	∞', (2, 3), (3, 3)	
	∞', (4, 1), (4, 2)	
	∞', (1, 1), (1, 2)	
	∞', ∞, (4, 3)	
	∞', (1, 2), (1, 3)	

3.3 *A Tight Circular DCCD*$(6n + 3, 3, 6n^2 + 5n + 1)$

Let $v = 6n + 3$ then $b_{\mathrm{MCT}}(6n + 3, 3) = b_e^c(6n + 3, 3) = 6n^2 + 5n + 1$.

For $n \geq 1$, we use the Bose construction of a STS$(6n + 3)$, from Section 1.2 of [9], and show that we can arrange its blocks to give a tight circular DCCD$(6n + 3, 3, 6n^2 + 5n + 1)$. Its padding $\mathcal{P}$ is the empty set.

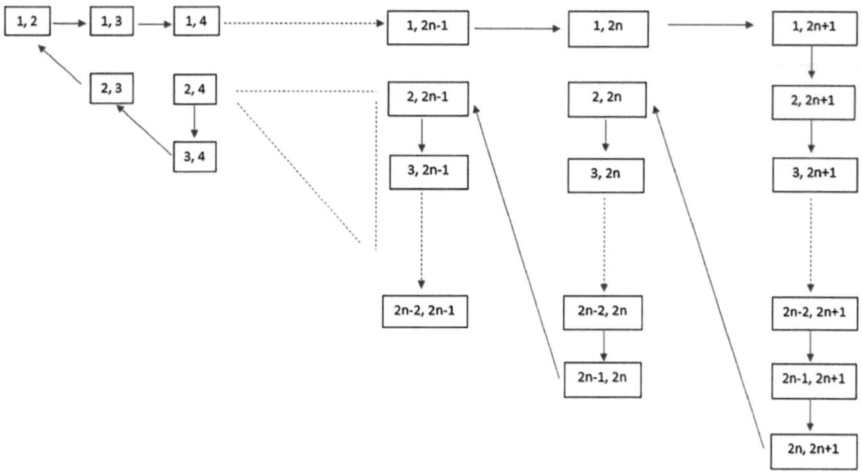

Fig. 6 Arrangement of $\boxed{i, j}$ for $\mathcal{B}_t$ for $t = 1, 2$, or 3

The Bose Construction Let $v = 6n + 3$ and let $(Q, \circ)$ be an idempotent commutative quasigroup of order $2n + 1$, where $Q = \{1, 2, \ldots, 2n + 1\}$. Let $S = Q \times \{1, 2, 3\}$, and define $\mathcal{T}$ to contain the following two types of blocks.

Type 1: $\{(i, 1), (i, 2), (i, 3)\}$ for $1 \le i \le 2n + 1$,
Type 2: $\{(i, 1), (j, 1), (i \circ j, 2)\}, \{(i, 2), (j, 2), (i \circ j, 3)\}, \{(i, 3), (j, 3), (i \circ j, 1)\}$ for $1 \le i < j \le 2n + 1$.

Then $(S, \mathcal{T})$ is a STS$(6n + 3)$.

We partition the blocks of Type 2. Let

- $\mathcal{B}_1$ be the blocks of Type 2 of the form $\{(i, 1), (j, 1), (i \circ j, 2)\}$ for $1 \le i < j \le 2n + 1$,
- $\mathcal{B}_2$ be the blocks of Type 2 of the form $\{(i, 2), (j, 2), (i \circ j, 3)\}$ for $1 \le i < j \le 2n + 1$,
- $\mathcal{B}_3$ be the blocks of Type 2 of the form $\{(i, 3), (j, 3), (i \circ j, 1)\}$ for $1 \le i < j \le 2n + 1$.

Similar to Fig. 2, now arrange them using pair i and j, i.e., $\boxed{i, j}$, as in Fig. 6.

The following Steps 1–4 construct a tight circular DCCD$(6n+3, 3, 6n^2+5n+1)$ for every $n \ge 1$. (Appropriately modified interpretations of some Steps are needed for $n = 1$.)

- **Step 1:** Let $Q = \{1, 2, \ldots, 2n + 1\}$. Consider the following idempotent commutative quasigroup of order $2n + 1$, $(Q, \circ)$, whose rows are formed from the back-circulant of $(1, n + 2, 2, n + 3, \ldots, i, n + i + 1, \ldots, 2n + 1, n + 1)$:

$\circ$	1	2	3	4	$\cdots$	$2n$	$2n+1$
1	1	$n+2$	2	$n+3$	$\cdots$	$2n+1$	$n+1$
2	$n+2$	2	$n+3$	3	$\cdots$	$n+1$	1
3	2	$n+3$	3	$n+4$	$\cdots$	1	$n+2$
4	$n+3$	3	$n+4$	4	$\cdots$	$n+2$	2
.	.	.	.	.	$\cdots$	.	.
.	.	.	.	.	$\cdots$	.	.
.	.	.	.	.	$\cdots$	.	.
$2n$	$2n+1$	$n+1$	1	$n+2$	$\cdots$	$2n$	n
$2n+1$	$n+1$	1	$n+2$	2	$\cdots$	n	$2n+1$

- **Step 2:** Going around the cycle in Fig. 6, arrange the blocks in $\mathcal{B}_t$ for $t = 1, 2,$ and 3 as follows: For $\mathcal{B}_1$, start with pair $\boxed{1, 3}$, follow the arrows, and finish with $\boxed{1, 2}$, i.e. start with block $\{(1, 1), (3, 1), (1 \circ 3, 2)\} = \{(1, 1), (3, 1), (2, 2)\}$ and finish with block $\{(1, 1), (2, 1), (1 \circ 2, 2)\} = \{(1, 1), (2, 1), (n + 2, 2)\}$. For both $\mathcal{B}_2$ and $\mathcal{B}_3$, start with pair $\boxed{1, 2}$ and finish with $\boxed{2, 3}$. See Fig. 7.
- **Step 3:** For $i = 2n + 1, 2n, 2n - 1, \ldots, 4$, insert the Type 1 blocks $\{(i, 1), (i, 2), (i, 3)\}$, shown in **bold**, into $\mathcal{B}_1$ as follows:

  for $i = 2n + 1$, block $\{(2n + 1,1),(2n + 1,2),(2n + 1,3)\}$ between $\{(1,1),(2n,1),(2n + 1,2)\}$ and $\{(1, 1), (2n + 1, 1), (n + 1, 2)\}$;

  for $i = 2n, 2n-1, \ldots, 4$, block $\{(i,1),(i,2),(i,3)\}$ between $\{(i,1),(i+1,1),(i \circ (i + 1),2)\}$ and $\{(2,1),(i,1),(2 \circ i,2)\}$.

  See Fig. 8a.

  It is straightforward to show that $i \circ (i + 1) = n + i + 1$ for $1 \le i \le 2n$, so $i \circ (i + 1) \ne i$. Also $2 \circ i = n + \frac{i+3}{2}$, $1 \le i \le 2n + 1$, i odd, and $2 \circ i = \frac{i+2}{2}$, $1 \le i \le 2n + 1$, i even, so $2 \circ i \ne i$ for $i = 2n, 2n - 1, \ldots, 4$. So these last block insertations will preserve the double-change between blocks.
- **Step 4:** From Fig. 8a, insert the following blocks, Type 1 blocks are shown in **bold**:

  At the end of $\mathcal{B}_1$ insert the Type 1 block $\{(1, 1), (1, 2), (1, 3)\}$,

  insert $\mathcal{B}_2$,

  insert the Type 1 block $\{(2, 1), (2, 2), (2, 3)\}$,

  insert $\mathcal{B}_3$,

  insert the Type 1 block $\{(3, 1), (3, 2), (3, 3)\}$.

  See Fig. 8b.

$(1, 1), (3, 1), \quad (2, 2)$	$(1, 2), (2, 2), (\boldsymbol{n} + 2, 3)$	$(1, 3), (2, 3), (\boldsymbol{n} + 2, 1)$
$(1, 1), (4, 1), (\boldsymbol{n} + 3, 2)$	$(1, 2), (3, 2), \quad (2, 3)$	$(1, 3), (3, 3), \quad (2, 1)$
. . .	. . .	. . .
. . .	. . .	. . .
. . .	. . .	. . .
$(1, 1), (2, 1), (\boldsymbol{n} + 2, 2)$	$(2, 2), (3, 2), (\boldsymbol{n} + 3, 3)$	$(2, 3), (3, 3), (\boldsymbol{n} + 3, 1)$
$\mathcal{B}_1$	$\mathcal{B}_2$	$\mathcal{B}_3$

Fig. 7 Blocks $\mathcal{B}_t$ for $t = 1, 2,$ and 3

$$\begin{array}{lll}
\infty', & (2,1), & (3,1)\\
\infty', & (2,2), & (3,2)\\
\infty', & (2,3), & (3,3)\\
\cdot & \cdot & \cdot\\
\cdot & \cdot & \cdot\\
\cdot & \cdot & \cdot\\
\infty', & (2i,1), & (2i+1,1)\\
\infty', & (2i,2), & (2i+1,2)\\
\infty', & (2i,3), & (2i+1,3)\\
\cdot & \cdot & \cdot\\
\cdot & \cdot & \cdot\\
\cdot & \cdot & \cdot\\
\infty', & (2n,1), & (2n+1,1)\\
\infty', & (2n,2), & (2n+1,2)\\
\infty', & (2n,3), & (2n+1,3)
\end{array}$$

(a)

$$\begin{array}{lll}
\infty', & (2,1), & (3,1)\\
\infty', & (2,2), & (3,2)\\
\infty', & (2,3), & (3,3)\\
\cdot & \cdot & \cdot\\
\cdot & \cdot & \cdot\\
\cdot & \cdot & \cdot\\
\infty', & (2i,1), & (2i+1,1)\\
\infty', & (2i,2), & (2i+1,2)\\
\infty', & (2i,3), & (2i+1,3)\\
\cdot & \cdot & \cdot\\
\cdot & \cdot & \cdot\\
\cdot & \cdot & \cdot\\
\infty', & (2n,1), & (2n+1,1)\\
\infty', & (2n,2), & (2n+1,2)\\
\boldsymbol{\infty'}, & \mathbf{(1,1)}, & \mathbf{(1,2)}\\
\infty', & (2n,3), & (2n+1,3)\\
\boldsymbol{\infty'}, & \mathbf{(1,2)}, & \mathbf{(1,3)}
\end{array}$$

(b)

Fig. 9 (**a**) Block arrangement for Step 5. (**b**) Block arrangement for Step 6

The Tripole Covering Construction Let $v \equiv 4 \pmod 6$, and $(X, \mathcal{T})$ be a STS of order $v - 1 \equiv 3 \pmod 6$. Let $\{a, b, c\} \in \mathcal{T}$ be any block in $\mathcal{T}$, and let $\phi = \{\{a, b\}, \{b, c\}\}$, and let $\pi = \{\{x_1, y_1\}, \{x_2, y_2\}, \ldots, \{x_{\frac{v-4}{2}}, y_{\frac{v-4}{2}}\}\}$ be any partition of $X \setminus \{a, b, c\}$. Now let $\phi(\infty') = \{\{\infty', a, b\}, \{\infty', b, c\}\}$, and $\pi(\infty') = \{\{\infty', x_1, y_1\}, \{\infty', x_2, y_2\}, \ldots, \{\infty', x_{\frac{v-4}{2}}, y_{\frac{v-4}{2}}\}\}$. Further, let $S = \{\infty'\} \cup X$, and $\mathcal{T}^* = \mathcal{T} \cup \phi(\infty') \cup \pi(\infty')$, and $\mathcal{P} = \phi \cup \pi \cup \{\{\infty', b\}\}$.

Then $(S, \mathcal{T}^*, \mathcal{P})$ is a MCT(v).

Let $(X, \mathcal{T})$ be the tight circular DCCD$(6n + 3, 3, 6n^2 + 5n + 1)$ of Fig. 8b, with $X = Q \times \{1, 2, 3\}$. Let $B_2 = \{(1, 1), (1, 2), (1, 3)\} \in \mathcal{T}$, then $\phi = \{\{(1, 1), (1, 2)\}, \{(1, 2), (1, 3)\}\}$. Let $\pi = \{\{(2i,1), (2i+1,1)\}, \{(2i,2), (2i+1,2)\}, \{(2i,3), (2i+1,3)\}$ $(1 \le i \le n)$ be a partition of $X \setminus \{(1, 1), (1, 2), (1, 3)\}$.

- **Steps 1–4** of Sect. 3.3.
- **Step 5:** Arrange the blocks of $\pi(\infty')$ in a double-change manner. See Fig. 9a.
- **Step 6:** Insert the 2 blocks in $\phi(\infty')$ shown in **bold** preserving the double-change, see Fig. 9b.
- **Step 7:** Insert the arrangement of blocks from Fig. 9b at the $\triangleleft$ in the tight circular DCCD$(6n + 3, 3, 6n^2 + 5n + 1)$ (after block $\{(2, 1), (2, 2), (2, 3)\}$) of Fig. 8b to give a minimum circular DCCD$(6n + 4, 3, 6n^2 + 8n + 3)$.

Now $\phi(\infty')$ contains 2 blocks, and π is a partition of $X \setminus \{(1, 1), (1, 2), (1, 3)\}$, a $6n$-set, into $3n$ pairs, so $\pi(\infty')$ contains $3n$ blocks. We check: $b_{\mathrm{MCT}}(6n + 3, 3) + 2 + 3n = 6n^2 + 5n + 1 + 3n + 2 = 6n^2 + 8n + 3 = b_{\mathrm{MCT}}(6n + 4, 3)$, as needed.

Example 6 $n = 2$, $v = 16$. We take the tight circular DCCD$(15, 3, 35)$ of Example 5 and add on one extra point ∞', and the $3n + 2 = 8$ new blocks $\phi(\infty') \cup \pi(\infty')$:

$$\begin{array}{l}\infty', (1,1), (1,2)\\ \infty', (1,2), (1,3)\\ \infty', (2,1), (3,1)\\ \infty', (2,2), (3,2)\\ \infty', (2,3), (3,3)\\ \infty', (4,1), (5,1)\\ \infty', (4,2), (5,2)\\ \infty', (4,3), (5,3)\end{array}$$

at the $\triangleleft$, to obtain the minimum circular DCCD(16, 3, 43):

(1, 1), (3, 1), (2, 2)	(1, 1), (1, 2), (1, 3)	(1, 3), (2, 3), (4, 1)
(1, 1), (4, 1), (5, 2)	(1, 2), (2, 2), (4, 3)	(1, 3), (3, 3), (2, 1)
(5, 1), (5, 2), (5, 3)	(1, 2), (3, 2), (2, 3)	(1, 3), (4, 3), (5, 1)
(1, 1), (5, 1), (3, 2)	(1, 2), (4, 2), (5, 3)	(1, 3), (5, 3), (3, 1)
(2, 1), (5, 1), (1, 2)	(1, 2), (5, 2), (3, 3)	(2, 3), (5, 3), (1, 1)
(3, 1), (5, 1), (4, 2)	(2, 2), (5, 2), (1, 3)	(3, 3), (5, 3), (4, 1)
(4, 1), (5, 1), (2, 2)	(3, 2), (5, 2), (4, 3)	(4, 3), (5, 3), (2, 1)
(4, 1), (4, 2), (4, 3)	(4, 2), (5, 2), (2, 3)	(2, 3), (4, 3), (3, 1)
(2, 1), (4, 1), (3, 2)	(2, 2), (4, 2), (3, 3)	(3, 3), (4, 3), (1, 1)
(3, 1), (4, 1), (1, 2)	(3, 2), (4, 2), (1, 3)	(2, 3), (3, 3), (5, 1)
(2, 1), (3, 1), (5, 2)	(2, 2), (3, 2), (5, 3)	(3, 1), (3, 2), (3, 3)
(1, 1), (2, 1), (4, 2)	(2, 1), (2, 2), (2, 3)	
	∞′, (2, 1), (3, 1)	
	∞′, (2, 2), (3, 2)	
	∞′, (2, 3), (3, 3)	
	∞′, (4, 1), (5, 1)	
	∞′, (4, 2), (5, 2)	
	∞′, (1, 1), (1, 2)	
	∞′, (4, 3), (5, 3)	
	∞′, (1, 2), (1, 3)	

3.5 An Economical Circular DCCD$(6n + 5, 3, 6n^2 + 9n + 4)$

Let $v = 6n + 5$ then $b_{\mathrm{MCT}}(6n + 5, 3) = 6n^2 + 9n + 4$.

From Sect. 1.1, in a DCCD(v, k) we have $v \geq k + 2$. Thus in a DCCD$(v, 3)$ we have $v \geq 5$, and so in Sects. 3.5 and 3.6 we also need to consider $n = 0$.

If $n = 0$, then $v = 5$ and $b_{\mathrm{MCT}}(5, 3) = 4$. Without loss of generality, let the 4 blocks in a MCT(5) be given as in Step 2 below. However these 4 blocks cannot be arranged in a linear nor in a circular double-change manner. It is straightforward to find a minimum circular DCCD(5, 3, 5).

For $n \geq 1$ see Section 1.4 of [9]:

Definition 4 A *pairwise balanced design* of order v, denoted by PBD(v), is an ordered pair $(S, \mathcal{B})$, where S is a finite set of v elements, and $\mathcal{B}$ is a collection of subsets of S called *blocks*, such that each pair of distinct elements of S occurs together in exactly one block of $\mathcal{B}$.

For all $n \geq 1$ we first need to construct a PBD($6n + 5$) with exactly one block of size 5 and the rest of size 3:

The 6*n* + 5 Construction Let $Q = \{1, 2, \ldots, 2n+1\}$ and $\alpha = (1)(23\cdots 2n+1)$. Let $(Q, \circ)$ be an idempotent commutative quasigroup of order $2n + 1$. Let $S = \{\infty_1, \infty_2\} \cup (\{1, 2, \ldots, 2n+1\} \times \{1, 2, 3\})$ and let $\mathcal{B}$ contain the following blocks:

Type 1: $\{\infty_1, \infty_2, (1, 1), (1, 2), (1, 3)\}$,
Type 2: $\{\infty_1, (2i, 1), (2i, 2)\}, \{\infty_1, (\alpha(2i),2), (\alpha(2i),3)\}, \{\infty_1, (2i,3), (\alpha(2i),1)\}$, $\{\infty_2, (\alpha(2i), 1), (\alpha(2i), 2)\}, \{\infty_2, (2i, 2), (2i, 3)\}, \{\infty_2, (\alpha^{-1}(2i), 3), (2i, 1)\}$ for $1 \leq i \leq n$,
Type 3: $\{(i,1), (j,1), (i \circ j,2)\}, \{(i,2), (j,2), (i \circ j,3)\}, \{(i,3), (j,3), (\alpha(i \circ j),1)\}$ for $1 \leq i < j \leq 2n+1$.
Then $(S, \mathcal{B})$ is a PBD($6n + 5$) with exactly one block of size 5 and the rest of size 3.

Next we use the double-edge covering construction of Section 4.3 of [9] to construct a MCT($6n + 5$). Its padding $\mathcal{P}$ is a double edge.

The Double-Edge Covering Construction Let $v \equiv 5 \pmod 6$ and let $(S, \mathcal{B})$ be a PBD(v) with one block $\{a, b, c, d, e\}$ of size 5 (i.e., the Type 1 block) and the rest of size 3. Let $\mathcal{T}^* = \{\{a, b, c\}, \{a, b, d\}, \{a, b, e\}, \{c, d, e\}\}$, and let $\mathcal{T}$ be the set of blocks of size 3, and $\mathcal{P} = \{\{a, b\}, \{a, b\}\}$.

Then $(S, \mathcal{T}^* \cup \mathcal{T}, \mathcal{P})$ is a MCT(v).

Steps 1–3 below construct an economical circular DCCD($6n+5, 3, 6n^2+9n+4$) for every $n \geq 1$.

- **Step 1:** Let $Q = \{1, 2, \ldots, 2n + 1\}$ and $\alpha = (1)(23\cdots 2n + 1)$. Let $(Q, \circ)$ be the idempotent commutative quasigroup of order $2n+1$ from Step 1 of Sect. 3.3. Construct the blocks of the above MCT($6n + 5$).
- **Step 2:** We have the 4 blocks $\mathcal{T}^*$ of size 3 from the Type 1 block:

  $\{\infty_1, \infty_2, (1, 1)\}$,
  $\{\infty_1, \infty_2, (1, 2)\}$,
  $\{\infty_1, \infty_2, (1, 3)\}$,
  $\{(1, 1), (1, 2), (1, 3)\}$.

  In the following, with slight abuse of notation, we also refer to these 4 blocks as "Type 1" blocks.

$(1,1), (2,1), (n+2,2)$	$(1,2), (2,2), (n+2,3)$	$(1,3), (3,3), \quad (\alpha(2),1)$
$(1,1), (3,1), \quad (2,2)$	$(1,2), (3,2), \quad (2,3)$	$(1,3), (4,3), \quad (\alpha(n+3),1)$
$\vdots$	$\vdots$	$\vdots$
$(2,1), (3,1), (n+3,2)$	$(2,2), (3,2), (n+3,3)$	$(1,3), (2,3), \quad (\alpha(n+2),1)$
$\mathcal{B}_1$	$\mathcal{B}_2$	$\mathcal{B}_3$

Fig. 10 Blocks $\mathcal{B}_t$ for $t = 1, 2$, and 3

We partition the blocks of Type 2 as follows:

Let $\mathcal{B}_{\infty_1}$ be the set of blocks of Type 2 that contain ∞_1, and arrange these blocks in the following double-change manner for $1 \le i \le n$:

$\{\infty_1, (2i, 1), (2i, 2)\}$,
$\{\infty_1, (\alpha(2i), 2), (\alpha(2i), 3)\}$,
$\{\infty_1, (2i, 3), (\alpha(2i), 1)\}$; and

let $\mathcal{B}_{\infty_2}$ be the set of the blocks of Type 2 that contain ∞_2, and arrange these in the following double-change manner for $1 \le i \le n$:

$\{\infty_2, (\alpha(2i), 1), (\alpha(2i), 2)\}$,
$\{\infty_2, (2i, 2), (2i, 3)\}$,
$\{\infty_2, (\alpha^{-1}(2i), 3), (2i, 1)\}$.

We partition the blocks of Type 3 by arranging them as in the Bose construction, see Step 2 of Sect. 3.3, with a small change: for both $\mathcal{B}_1$ and $\mathcal{B}_2$, start with pair $\boxed{1, 2}$, follow the arrows, and finish with $\boxed{2, 3}$, and for $\mathcal{B}_3$, start with pair $\boxed{1, 3}$ and finish with $\boxed{1, 2}$. See Fig. 10.

- **Step 3:**

  Start with the Type 1 block $\{\infty_1, \infty_2, (1, 1)\}$,
  insert $\mathcal{B}_1$,
  insert $\mathcal{B}_{\infty_1}$,
  insert the Type 1 block $\{\infty_1, \infty_2, (1, 2)\}$,
  insert $\mathcal{B}_2$,
  insert $\mathcal{B}_{\infty_2}$,
  insert the Type 1 block $\{\infty_1, \infty_2, (1, 3)\}$,
  insert $\mathcal{B}_3$,
  insert the Type 1 block $\{(1, 1), (1, 2), (1, 3)\}$.

  See Fig. 11, where, in Fig. 11b, for clarity, the 4 Type 1 blocks are shown in **bold.**

We check that Fig. 11b is an economical circular DCCD$(6n+5, 3, 6n^2+9n+4)$ for every $n \ge 1$.

(a)

$\infty_1,\ \infty_2,\ (1,1)$
$\mathcal{B}_1$
$\mathcal{B}_{\infty_1}$
$\infty_1,\ \infty_2,\ (1,2)$
$\mathcal{B}_2$
$\mathcal{B}_{\infty_2}$
$\infty_1,\ \infty_2,\ (1,3)$
$\mathcal{B}_3$
$(1,1), (1,2), (1,3)$

(b)

$\boldsymbol{\infty_1,\ \infty_2,\ (1,1)}$
$(1,1), (2,1), (n+2,2)$
$(1,1), (3,1), (2,2)$
$\vdots$
$(2,1), (3,1), (n+3,2)$
$\infty_1,\ (2,1),\ (2,2)$
$\infty_1, (\alpha(2),2), (\alpha(2),3)$
$\vdots$
$\infty_1,\ (2n,3),\ (\alpha(2n),1)$
$\boldsymbol{\infty_1,\ \infty_2,\ (1,2)}$
$(1,2),\ (2,2),\ (n+2,3)$
$(1,2),\ (3,2),\ (2,3)$
$\vdots$
$(2,2),\ (3,2),\ (n+3,3)$
$\infty_2,\ (\alpha(2),1), (\alpha(2),2)$
$\infty_2,\ (2,2),\ (2,3)$
$\vdots$
$\infty_2, (\alpha^{-1}(2n),3), (2n,1)$
$\boldsymbol{\infty_1,\ \infty_2,\ (1,3)}$
$(1,3),\ (3,3),\ (\alpha(2),1)$
$(1,3),\ (4,3),\ (\alpha(n+3),1)$
$\vdots$
$(1,3),\ (2,3),\ (\alpha(n+2),1)$
$\boldsymbol{(1,1),\ (1,2),\ (1,3)}$

Fig. 11 (**a**) Block arrangement for Step 3. (**b**) Detailed block arrangement for Step 3. An economical circular DCCD$(6n+5, 3, 6n^2+9n+4)$

In this construction we have 4 Type 1 blocks, $6n$ blocks of Type 2, and $3n(2n+1)$ blocks of Type 3. Thus the number of blocks $b = 6n^2 + 9n + 4 = b_{\text{MCT}}(6n+5, 3)$, as needed.

Example 7 $n = 2, v = 17, Q = \{1, 2, 3, 4, 5\}$. We use the idempotent commutative quasigroup of order 5, $(Q, \circ)$, from Example 5 of Sect. 3.3, to construct an economical circular DCCD$(17, 3, 46)$:

$\infty_1, \infty_2, (1,1)$	$\infty_1, \infty_2, (1,2)$	$\infty_1, \infty_2, (1,3)$
$(1,1),(2,1),(4,2)$	$(1,2),(2,2),(4,3)$	$(1,3),(3,3),(3,1)$
$(1,1),(3,1),(2,2)$	$(1,2),(3,2),(2,3)$	$(1,3),(4,3),(2,1)$
$(1,1),(4,1),(5,2)$	$(1,2),(4,2),(5,3)$	$(1,3),(5,3),(4,1)$
$(1,1),(5,1),(3,2)$	$(1,2),(5,2),(3,3)$	$(2,3),(5,3),(1,1)$
$(2,1),(5,1),(1,2)$	$(2,2),(5,2),(1,3)$	$(3,3),(5,3),(5,1)$
$(3,1),(5,1),(4,2)$	$(3,2),(5,2),(4,3)$	$(4,3),(5,3),(3,1)$
$(4,1),(5,1),(2,2)$	$(4,2),(5,2),(2,3)$	$(2,3),(4,3),(4,1)$
$(2,1),(4,1),(3,2)$	$(2,2),(4,2),(3,3)$	$(3,3),(4,3),(1,1)$
$(3,1),(4,1),(1,2)$	$(3,2),(4,2),(1,3)$	$(2,3),(3,3),(2,1)$
$(2,1),(3,1),(5,2)$	$(2,2),(3,2),(5,3)$	$(1,3),(2,3),(5,1)$
$\infty_1, (2,1),(2,2)$	$\infty_2, (3,1),(3,2)$	$(1,1),(1,2),(1,3)$
$\infty_1, (3,2),(3,3)$	$\infty_2, (2,2),(2,3)$	
$\infty_1, (2,3),(3,1)$	$\infty_2, (5,3),(2,1)$	
$\infty_1, (4,1),(4,2)$	$\infty_2, (5,1),(5,2)$	
$\infty_1, (5,2),(5,3)$	$\infty_2, (4,2),(4,3)$	
$\infty_1, (4,3),(5,1)$	$\infty_2, (3,3),(4,1)$	

3.6 A Minimum Circular DCCD$(6n + 6, 3, 6n^2 + 12n + 6)$

Let $v = 6n + 6$ then $b_{\mathrm{MCT}}(6n + 6, 3) = 6n^2 + 12n + 6$.

If $n = 0$ then $v = 6$ and $b_{\mathrm{MCT}}(6, 3) = 6$, and the 6 blocks in Step 2 below form a MCT(6) and these blocks can be arranged in a circular double-change manner, to give a minimum circular DCCD(6, 3, 6).

For $n \geq 1$ we use the $6n + 5$ construction of Sect. 3.5 and then the 1-factor covering construction of Section 4.3 of [9] to construct a MCT$(6n + 6)$. Its padding is a 1-factor of K_{6n+6}, i.e., $\mathcal{P} = (3n + 3)K_2$.

The 1-Factor Covering Construction Let $v \equiv 0 \pmod 6$ and $(X, \mathcal{B})$ be a PBD$(v - 1)$ of order $v - 1 \equiv 5 \pmod 6$ with one block $\{a, b, c, d, e\}$ of size 5, and the rest of size 3. Let $\mathcal{T}$ be the collection of blocks of size 3. Let $S = \{\infty\} \cup X$ and $\mathcal{F}(\infty) = \{\{\infty, a, e\}, \{\infty, b, e\}, \{\infty, c, d\}, \{a, b, c\}, \{a, b, d\}, \{c, d, e\}\}$. Let $\pi = \{\{x_1, y_1\}, \{x_2, y_2\}, \ldots, \{x_{\frac{v-6}{2}}, y_{\frac{v-6}{2}}\}\}$ be any partition of $X \setminus \{a, b, c, d, e\}$, and let $\pi(\infty) = \{\{\infty, x_1, y_1\}, \{\infty, x_2, y_2\}, \ldots, \{\infty, x_{\frac{v-6}{2}}, y_{\frac{v-6}{2}}\}\}$. Finally, let $\mathcal{T}^* = \mathcal{F}(\infty) \cup \pi(\infty) \cup \mathcal{T}$, and $\mathcal{P} = \pi \cup \{\{a, b\}, \{c, d\}, \{e, \infty\}\}$.

Then $(S, \mathcal{T}^*, \mathcal{P})$ is a MCT(v).

Steps 1–3 below construct a minimum circular DCCD$(6n + 6, 3, 6n^2 + 12n + 6)$ for every $n \geq 1$.

- **Step 1:** We use the $6n + 5$ construction of Sect. 3.5 along with the 1-factor covering construction described above to obtain the blocks of a MCT$(6n + 6)$.
- **Step 2:** From the Type 1 block we form the following 6 blocks in $\mathcal{F}(\infty)$:

  $\{\infty, \infty_1, (1, 3)\}$,
  $\{\infty, \infty_2, (1, 3)\}$,
  $\{\infty, (1, 1), (1, 2)\}$,
  $\{\infty_1, \infty_2, (1, 1)\}$,
  $\{\infty_1, \infty_2, (1, 2)\}$,
  $\{(1, 1), (1, 2), (1, 3)\}$.

  In the following, with slight abuse of notation, we also refer to these 6 blocks as "Type 1" blocks.

  Now, because π partitions the set $X \setminus \{\infty_1, \infty_2, (1, 1), (1, 2), (1, 3)\}$, then the blocks of $\pi(\infty)$ have a double-change with ∞ as the common element. We keep the blocks in $\mathcal{T}$, i.e., in $\mathcal{B}_{\infty_1}$, $\mathcal{B}_{\infty_2}$, $\mathcal{B}_1$, $\mathcal{B}_2$, and $\mathcal{B}_3$, the same as in Step 2 of Sect. 3.5.
- **Step 3:**

  Start with the Type 1 block $\{\infty_1, \infty_2, (1, 1)\}$,
  insert $\mathcal{B}_1$,
  insert $\mathcal{B}_{\infty_1}$,
  insert the Type 1 block $\{\infty_1, \infty_2, (1, 2)\}$,
  insert $\mathcal{B}_2$,
  insert $\mathcal{B}_{\infty_2}$,
  insert the Type 1 block $\{\infty, \infty_2, (1, 3)\}$,
  insert the Type 1 block $\{\infty, (1, 1), (1, 2)\}$,
  insert $\pi(\infty)$,
  insert the Type 1 block $\{\infty, \infty_1, (1, 3)\}$,
  insert $\mathcal{B}_3$,
  insert the Type 1 block $\{(1, 1), (1, 2), (1, 3)\}$.

  See Fig. 12, where, in Fig. 12b, for clarity, the 6 Type 1 blocks are shown in **bold.**

We check that Fig. 12b is a minimum circular DCCD$(6n + 6, 3, 6n^2 + 12n + 6)$ for every $n \geq 1$.

In this construction we have 6 Type 1 blocks, $3n$ blocks in $\pi(\infty)$, $6n$ blocks of Type 2 (from the $6n + 5$ construction of Sect. 3.5), and $3n(2n + 1)$ blocks of Type 3 (from the $6n + 5$ construction of Sect. 3.5). Thus the number of blocks $b = 6n^2 + 12n + 6 = b_{\text{MCT}}(6n + 6, 3)$, as needed.

$$
\begin{array}{ccc}
\infty_1, & \infty_2, & (1,1) \\
 & \mathcal{B}_1 & \\
 & \mathcal{B}_{\infty_1} & \\
\infty_1, & \infty_2, & (1,2) \\
 & \mathcal{B}_2 & \\
 & \mathcal{B}_{\infty_2} & \\
\infty, & \infty_2, & (1,3) \\
\infty, & (1,1), & (1,2) \\
 & \pi(\infty) & \\
\infty, & \infty_1, & (1,3) \\
 & \mathcal{B}_3 & \\
(1,1), & (1,2), & (1,3)
\end{array}
$$

(a)

$$
\begin{array}{ccc}
\boldsymbol{\infty_1}, & \boldsymbol{\infty_2}, & \mathbf{(1,1)} \\
(1,1), & (2,1), & (n+2,2) \\
(1,1), & (3,1), & (2,2) \\
\cdot & \cdot & \cdot \\
\cdot & \cdot & \cdot \\
\cdot & \cdot & \cdot \\
(2,1), & (3,1), & (n+3,2) \\
\infty_1, & (2,1), & (2,2) \\
\infty_1, & (\alpha(2),2), & (\alpha(2),3) \\
\cdot & \cdot & \cdot \\
\cdot & \cdot & \cdot \\
\cdot & \cdot & \cdot \\
\infty_1, & (2n,3), & (\alpha(2n),1) \\
\boldsymbol{\infty_1}, & \boldsymbol{\infty_2}, & \mathbf{(1,2)} \\
(1,2), & (2,2), & (n+2,3) \\
(1,2), & (3,2), & (2,3) \\
\cdot & \cdot & \cdot \\
\cdot & \cdot & \cdot \\
\cdot & \cdot & \cdot \\
(2,2), & (3,2), & (n+3,3) \\
\infty_2, & (\alpha(2),1), & (\alpha(2),2) \\
\infty_2, & (2,2), & (2,3) \\
\cdot & \cdot & \cdot \\
\cdot & \cdot & \cdot \\
\cdot & \cdot & \cdot \\
\infty_2, & (\alpha^{-1}(2n),3), & (2n,1) \\
\boldsymbol{\infty}, & \boldsymbol{\infty_2}, & \mathbf{(1,3)} \\
\boldsymbol{\infty}, & \mathbf{(1,1)}, & \mathbf{(1,2)} \\
\infty, & (2,1), & (2,2) \\
\infty, & (2,3), & (3,1) \\
\cdot & \cdot & \cdot \\
\cdot & \cdot & \cdot \\
\cdot & \cdot & \cdot \\
\infty, & (2n+1,2), & (2n+1,3) \\
\boldsymbol{\infty}, & \boldsymbol{\infty_1}, & \mathbf{(1,3)} \\
(1,3), & (3,3), & (\alpha(2),1) \\
(1,3), & (4,3), & (\alpha(n+3),1) \\
\cdot & \cdot & \cdot \\
\cdot & \cdot & \cdot \\
\cdot & \cdot & \cdot \\
(1,3), & (2,3), & (\alpha(n+2),1) \\
\mathbf{(1,1)}, & \mathbf{(1,2)}, & \mathbf{(1,3)}
\end{array}
$$

(b)

Fig. 12 (**a**) Block arrangement for Step 3. (**b**) Detailed block arrangement for Step 3. A minimum circular DCCD$(6n + 6, 3, 6n^2 + 12n + 6)$

Example 8 $n = 2, v = 18, Q = \{1, 2, 3, 4, 5\}$. We use the idempotent commutative quasigroup of order 5, $(Q, \circ)$, from Example 5 of Sect. 3.3, and construct a minimum circular DCCD(18, 3, 54):

$\infty_1, \ \infty_2, \ (1,1)$	$\infty_1, \ \infty_2, \ (1,2)$	$\infty, \ \infty_2, \ (1,3)$
$(1,1), (2,1), (4,2)$	$(1,2), (2,2), (4,3)$	$\infty, \ (1,1), (1,2)$
$(1,1), (3,1), (2,2)$	$(1,2), (3,2), (2,3)$	$\infty, \ (2,1), (2,2)$
$(1,1), (4,1), (5,2)$	$(1,2), (4,2), (5,3)$	$\infty, \ (2,3), (3,1)$
$(1,1), (5,1), (3,2)$	$(1,2), (5,2), (3,3)$	$\infty, \ (3,2), (3,3)$
$(2,1), (5,1), (1,2)$	$(2,2), (5,2), (1,3)$	$\infty, \ (4,1), (4,2)$
$(3,1), (5,1), (4,2)$	$(3,2), (5,2), (4,3)$	$\infty, \ (4,3), (5,1)$
$(4,1), (5,1), (2,2)$	$(4,2), (5,2), (2,3)$	$\infty, \ (5,2), (5,3)$
$(2,1), (4,1), (3,2)$	$(2,2), (4,2), (3,3)$	$\infty, \ \infty_1, \ (1,3)$
$(3,1), (4,1), (1,2)$	$(3,2), (4,2), (1,3)$	$(1,3), (3,3), (3,1)$
$(2,1), (3,1), (5,2)$	$(2,2), (3,2), (5,3)$	$(1,3), (4,3), (2,1)$
$\infty_1, \ (2,1), (2,2)$	$\infty_2, \ (3,1), (3,2)$	$(1,3), (5,3), (4,1)$
$\infty_1, \ (3,2), (3,3)$	$\infty_2, \ (2,2), (2,3)$	$(2,3), (5,3), (1,1)$
$\infty_1, \ (2,3), (3,1)$	$\infty_2, \ (5,3), (2,1)$	$(3,3), (5,3), (5,1)$
$\infty_1, \ (4,1), (4,2)$	$\infty_2, \ (5,1), (5,2)$	$(4,3), (5,3), (3,1)$
$\infty_1, \ (5,2), (5,3)$	$\infty_2, \ (4,2), (4,3)$	$(2,3), (4,3), (4,1)$
$\infty_1, \ (4,3), (5,1)$	$\infty_2, \ (3,3), (4,1)$	$(3,3), (4,3), (1,1)$
		$(2,3), (3,3), (2,1)$
		$(1,3), (2,3), (5,1)$
		$(1,1), (1,2), (1,3)$

4 Regular Minimum DCCD(v, k, b) and Regular Tight DCCD(v, k, b) with $k = 2$ or 3

In later papers we will study regular DCCD(v, k, b) for $k \geq 4$. For completeness we consider $k = 2$ or 3 in this paper.

Definition 5 A DCCD(v, k, b) $\mathcal{D}$ is *regular* if each element $x \in [v]$ is contained in the same number of blocks of $\mathcal{D}$. If each x is contained in r blocks of $\mathcal{D}$, we say that $\mathcal{D}$ is r*-regular*.

For example, the DCCD(7, 3, 7) in Example 1 is 3-regular.

In an r-regular DCCD(v, k, b) we have $vr = kb$.

We observe that in a DCCD(v, k, b) each element $x \in [v]$ must be paired with the $v - 1$ elements in $[v] \setminus \{x\}$.

We determine when minimum DCCD$(v, 2, b)$ and DCCD$(v, 3, b)$ are regular:

DCCD(v, 2, b) From Sect. 2 we have:

Theorem 4

(i) For $v = 5$ every minimum linear DCCD$(5, 2, 10)$ is 4-regular and tight, and every minimum circular DCCD$(5, 2, 11)$ is neither regular nor tight.
(ii) For $v \geq 6$ every minimum linear or circular DCCD$(v, 2, \binom{v}{2})$ is $(v-1)$-regular and tight.

□

DCCD$(v, 3, b)$ From Sect. 3 and the paddings $\mathcal{P}$ for each of the 6 cases of $v \pmod 6$ we have for both linear and circular DCCD$(v, 3, b)$'s:

Theorem 5

(i) For $v \equiv 1 \pmod 6$ i.e., $v = 6n+1$, every minimum DCCD$(6n+1, 3, 6n^2+n)$ is $3n$-regular and tight.
(ii) For $v \equiv 3 \pmod 6$ i.e., $v = 6n + 3$, every minimum DCCD$(6n + 3, 3, 6n^2 + 5n + 1)$ is $(3n + 1)$-regular and tight.
(iii) For $v \equiv 6 \pmod 6$ i.e., $v = 6n + 6$, every minimum DCCD$(6n + 6, 3, 6n^2 + 12n + 6)$ is $(3n + 3)$-regular and not tight.
(iv) For $v \equiv 2, 4,$ or $5 \pmod 6$ every minimum DCCD$(v, 3, b)$ is neither regular nor tight.

□

Acknowledgments We would like to thank the referees whose comments much improved the clarity of this paper.

References

1. A. Chafee, Single and double-change covering designs. MSc. Thesis, Carleton University, Ottawa (2021)
2. A. Chafee, B. Stevens, Linear and circular single-change covering designs re-visited (2022). arXiv:2209.11010. [math CO]
3. C.J. Colbourn, J.H. Dinitz, *Handbook of Combinatorial Designs*, 2nd edn. (Chapman and Hall/CRC Press, Boca Raton, 2007)
4. Y. Dimitrov, Double-change covering designs (in Bulgarian). Technical Report, Sofia University (1997)
5. M.K. Fort, Jr., G.A. Hedlund, Minimal coverings of pairs by triples. Paci. J. Math. **8**(4), 709–719 (1958)
6. N.T. Gamachchige, Double-change circular covering designs: constructions and applications. MSc Research Paper, Southern Illinois University (2014)
7. P. Horak, A. Rosa, Decomposing Steiner triple systems into small configurations. Ars Combin. **26** 91–105 (1988)
8. K.A. Lindbloom, Tight double-change covering designs. MSc Research Paper, Southern Illinois University (2011)
9. C.C. Lindner, C.A. Rodger, *Design Theory*, 2nd edn. (CRC Press, Boca Raton, 2009)
10. D.A. Preece, R.L. Constable, G. Zhang, J.L. Yucas, W.D. Wallis, J.P. McSorley, N.C.K. Phillips, Tight single-change covering designs. Utilitas Math. **47**, 55–84 (1995)
11. W.D. Wallis, J.L. Yucas, G.-H. Zhang, Single-change covering designs. Des. Codes Cryptograph. **3**, 9–19 (1992)

More formally, let us consider the set of formal power series ring $\mathcal{F} = \mathbb{K}[[t]]$, where $\mathbb{K}$ is the field $\mathbb{R}$ or $\mathbb{C}$. The *order* of $f(t) \in \mathcal{F}$, $f(t) = \sum_{k=0}^{\infty} f_k t^k$ ($f_k \in \mathbb{K}$), is the minimum number $r \in \mathbb{N}$ such that $f_r \neq 0$, where $\mathbb{N} = \{0, 1, 2 \ldots\}$ is the set of all natural numbers. We denote by $\mathcal{F}_r$ the set of formal power series of order r. Let $g(t) \in \mathcal{F}_0$ and $f(t) \in \mathcal{F}_1$; the pair (g, f) defines the *(proper) Riordan array* $D = (d_{n,k})_{n,k\in\mathbb{N}} = (g, f)$ having

$$d_{n,k} = [t^n]g(t)f(t)^k \tag{1}$$

or, in other words, having gf^k as the generating function of the kth column of (g, f). The *first fundamental theorem of Riordan arrays* can be represented as

$$(g(t), f(t))h(t) = g(t)(h \circ f)(t),$$

which can also be simplified to $(g, f)h = gh(f)$. Thus we immediately see that the usual row-by-column product of two Riordan arrays is also a Riordan array:

$$(g_1, f_1)(g_2, f_2) = (g_1 g_2(f_1), f_2(f_1)). \tag{2}$$

The Riordan array $I = (1, t)$ is the identity matrix because its entries are $d_{n,k} = [t^n]t^k = \delta_{n,k}$.

Let $(g(t), f(t))$ be a Riordan array. Then its inverse is

$$(g(t), f(t))^{-1} = \left(\frac{1}{g(\overline{f}(t))}, \overline{f}(t), \right) \tag{3}$$

where $\overline{f}(t)$ is the compositional inverse of $f(t)$, i.e., $(f \circ \overline{f})(t) = (\overline{f} \circ f)(t) = t$. In this way, the set $\mathcal{R}$ of all proper Riordan arrays forms a group (see [12]) called the Riordan group.

Here is a list of six important subgroups of the Riordan group (see [1, 11–13]).

- the *Appell subgroup* $\mathcal{A} = \{(g(t), t) : g \in \mathcal{F}_0\}$.
- the *Lagrange (associated) subgroup* $\mathcal{L} = \{(1, f(t) : f \in \mathcal{F}_1)\}$.
- the *k-Bell subgroup* $\mathcal{B}_k = \{(g(t), t(g(t))^k) : g \in \mathcal{F}_0\}$, where k is a fixed positive integer.
- the *hitting-time subgroup* $\mathcal{H} = \{(tf'(t)/f(t), f(t)) : f \in \mathcal{F}_1\}$.
- the *derivative subgroup* $\mathcal{D} = \{(f'(t), f(t)) : f \in \mathcal{F}_1\}$.
- the *checkerboard subgroup* $C = \{(g(t), f(t)) : g \in \mathcal{F}_0, f \in \mathcal{F}_1\}$, where g is an even function and f is an odd function.

The 1-Bell subgroup $\mathcal{B}_1$ is referred to as the Bell subgroup and denoted by $\mathcal{B}$ for short, and the Appell subgroup $\mathcal{A}$ can be considered as the 0-Bell subgroup $\mathcal{B}_0$ if we allow $k = 0$ to be included in the definition of the k-Bell subgroup $\mathcal{B}_k$.

An infinite lower triangular matrix $[d_{n,k}]_{n,k\in\mathbb{N}}$ is a Riordan array if and only if a unique sequence $A = (a_0 \neq 0, a_1, a_2, \ldots)$ exists such that for every $n, k \in \mathbb{N}$

$$d_{n+1,k+1} = a_0 d_{n,k} + a_1 d_{n,k+1} + \cdots + a_n d_{n,n}. \tag{4}$$

This is equivalent to

$$f(t) = tA(f(t)) \quad \text{or} \quad t = \bar{f}(t)A(t). \tag{5}$$

Here, $A(t)$ is the generating function of the A-sequence. The first formula of (5) is also called the *second fundamental theorem of Riordan arrays*. Moreover, there exists a unique sequence $Z = (z_0, z_1, z_2, \ldots)$ such that every element in column 0 can be expressed as the linear combination

$$d_{n+1,0} = z_0 d_{n,0} + z_1 d_{n,1} + \cdots + z_n d_{n,n}, \tag{6}$$

or equivalently,

$$g(t) = \frac{1}{1 - tZ(f(t))}, \tag{7}$$

in which and thoroughly we always assume $g(0) = g_0 = 1$, a usual hypothesis for proper Riordan arrays. From (7), we may obtain

$$Z(t) = \frac{g(\bar{f}(t)) - 1}{\bar{f}(t)g(\bar{f}(t))}.$$

A- and Z-sequence characterizations of Riordan arrays were introduced, developed, and/or studied in Merlini et al. [9], Roger [10], Sprugnoli and the author [4], Jean-Louis and Nkwanta [5], Luzón et al. [6, 7], and [2] as well as their references. In [4] the expressions of the A- and Z-sequences of the product depend on the analogous sequences of the two factors are given. More precisely, we consider two proper Riordan matrices $D_1 = (g_1, f_1)$ and $D_2 = (g_2, f_2)$ and their product,

$$D_3 = D_1 D_2 = (g_1 g_2(f_1), f_2(f_1)).$$

Denote by $A_i(t)$ and $Z_i(t)$, $i = 1, 2$, and 3, the generating functions of A-sequences and Z-sequences of D_i, $i = 1, 2$, and 3, respectively. Then

$$A_3(t) = A_2(t)A_1\left(\frac{t}{A_2(t)}\right) \tag{8}$$

and

$$Z_3(t) = \left(1 - \frac{t}{A_2(t)} Z_2(t)\right) Z_1\left(\frac{t}{A_2(t)}\right) + A_1\left(\frac{t}{A_2(t)}\right) Z_2(t). \tag{9}$$

Let $A(t)$ and $Z(t)$ be the generating functions of the A- and Z-sequences of a Riordan matrix $D = (g, f)$, and let $A^*(t)$ and $Z^*(t)$ be the generating functions of the A- and Z-sequences of the inverse Riordan matrix $D^{-1} = (1/g(\overline{f}), \overline{f})$. We immediately observe that

$$A^*\left(\frac{t}{A(t)}\right) = \frac{1}{A(t)} \tag{10}$$

and

$$Z^*\left(\frac{t}{A(t)}\right) = \frac{Z(t)}{tZ(t) - A(t)}. \tag{11}$$

In a Riordan array $(g, f) = (d_{n,k})_{n,k\geq 0}$, the first column (0th column), with its generating function $g(t)$, usually possesses an interesting combinatorial interpretation or represents an important sequence, while the other columns might be considered as the compositions of the first column in the view of the following recurrence relation:

$$d_{n,k} = [t^n]gf^k = \sum_{j=1}^{n} f_j[t^{n-j}]gf^{k-1} = \sum_{j=1}^{n-k+1} f_j d_{n-j,k-1} \tag{12}$$

for $n, k \geq 1$, where $f(t) = \sum_{j\geq 1} f_j t^j$. Equation (12) is called the *vertical recurrence relation* of (g, f). For instance,

$$\begin{aligned}
&d_{1,1} = f_1 d_{0,0},\\
&d_{2.1} = f_2 d_{0,0} + f_1 d_{1,0},\ d_{2,2} = f_1 d_{1,1},\\
&d_{3,1} = f_3 d_{0,0} + f_2 d_{1,0} + f_1 d_{2.0},\ d_{3,2} = f_2 d_{2,1} + f_1 d_{1,1},\ d_{3,3} = f_1 d_{2,2}, \ldots
\end{aligned}$$

Let $(g, f) = (d_{n,k})_{n,k\geq 0}$ be a Riordan array. Then its entries $d_{n,k}$, $n, k \geq 0$, can be represented recursively by (cf. [3])

$$\begin{aligned}
d_{n,k} = &\sum_{i_k=1}^{n-(k-1)} f_{i_k} \sum_{i_{k-1}=1}^{n-(k-2)-i_k} f_{i_{k-1}} \sum_{i_{k-2}=1}^{n-(k-3)-i_k-i_{k-1}} f_{i_{k-2}} \cdots\\
&\sum_{i_1=1}^{n-i_k-i_{k-1}-i_2} f_{i_1} g_{n-i_1-i_2-\cdots-i_k}.
\end{aligned} \tag{13}$$

Definition 1 (cf. [3]) Let $g \in \mathcal{F}_0$ with $g(0) = 1$ and $f \in \mathcal{F}_1$. We let $[g, f]$ denote the following array, called a quasi-Riordan array.

$$[g, f] := (g, f, tf, t^2 f, \ldots), \tag{14}$$

where $g, f, tf, t^2f \cdots$ are the generating functions of the 0th, 1st, 2nd, 3rd, $\cdots$, columns of the matrix $[g, f]$, respectively.

Let $\mathcal{R}_r$ denote the set of all quasi-Riordan arrays defined by (14). From [3], The set $\mathcal{R}_r$ is a group with the identity $[1, t]$, called the quasi-Riordan group, with respect to the multiplication defined by

$$[g, f][d, h] = \left[g + \frac{f}{t}(d-1), \frac{fh}{t}\right]. \tag{15}$$

By (15), the inverse of $[g, f]$ is

$$[g, f]^{-1} = \left[1 + \frac{t}{f}(1-g), \frac{t^2}{f}\right]. \tag{16}$$

The vertical recurrence relation of the column entries of Riordan arrays has many applications. For example, Mao et al. [8] use the relation to give an interesting criterion for the total positivity of Riordan arrays.

In next section, we study the quasi-Riordan group, the set of all the matrices represented the vertical recurrence relations of Riordan arrays. More precisely, we give the normal subgroup of the quasi-Riordan group and discuss the overlap of the quasi-Riordan and the Riordan group, the Appell subgroup. By the discussion, several extended subgroups of the quasi-Riordan group and the Riordan group are obtained. The horizontal recurrence relation and vertical recurrence relation are given accordingly. Finally, we show that the A-sequences and the vertical recurrence relations of the Riordan subgroups and those of the corresponding extensions are the same.

2 Main Results

Let $[g, f]$ be a quasi-Riordan array defined by (14). We now consider when it is also a Riordan array denoted by (d, h). Then from the definitions of quasi-Riordan array and Riordan array, we have

$$[g, f] = (g, f, tf, t^2f, \cdots) \quad \text{and} \quad (d, h) = (d, dh, dh^2, dh^3, \cdots).$$

Thus the quasi-Riordan array $[g, f]$ is a Riordan array (d, h) if and only if

$$(g, f, tf, t^2f, \cdots) = (d, dh, dh^2, dh^3, \cdots),$$

i.e., if and only if $g = d, h = t$, and $f = tg$, which implies $[g, f] = [g, tg] = (g, t) \in \mathcal{A}$. Where $\mathcal{A}$ is the Appell subgroup of the Riordan group.

of which first few entries are

$$\left[\frac{C(t)-1}{t}, tC(t)\right] = \left(\frac{C(t)-1}{t}, tC(t), t^2C(t), \ldots\right) = \begin{bmatrix} 1 & 0 & 0 & 0 & 0 & 0 & \cdots \\ 2 & 1 & 0 & 0 & 0 & 0 & \\ 5 & 1 & 1 & 0 & 0 & 0 & \\ 14 & 2 & 1 & 1 & 0 & 0 & \\ 42 & 5 & 2 & 1 & 1 & 0 & \\ 132 & 14 & 5 & 2 & 1 & 1 & \\ & & \cdots & & & & \ddots \end{bmatrix}.$$

Similarly, we have the following extensions of subgroups of the Riordan group.

Theorem 2 *Let $\mathcal{R}$ be the Riordan group, and let $\mathcal{A}$, $\mathcal{L}$, $\mathcal{B}_k$, $\mathcal{H}$, $\mathcal{D}$, and C be the Appell, Lagrange, k-Bell, hitting-time, derivative, and checkerboard subgroups of $\mathcal{R}$ defined before. Then, $\mathcal{LA}$, $\mathcal{B}_k\mathcal{A}$, $\mathcal{HA}$, $\mathcal{DA}$, and $C\mathcal{A}$ are also the subgroups of $\mathcal{R}$, which are called the extended Lagrange, the extended k-Bell, the extended hitting-time, the extended derivative, and the extended checkerboard subgroups of $\mathcal{R}$, respectively, where $\mathcal{LA} = \mathcal{R}$. The extended Appell subgroup is itself.*

Finally, we discuss the horizontal recurrence relation and vertical recurrence relation of the extended subgroups of the Riordan group.

Proposition 4 *Denote by $(d_{n,k})_{n,k\geq 0}$ and $(e_{n,k})_{n,k\geq 0}$ an element of a subgroup M of the Riordan group and the corresponding element of the corresponding extended subgroup $M\mathcal{A}$. Then the horizontal recurrence relations of the entries $d_{n,k}$ and the entries $e_{n,k}$ for $k \geq 1$ are the same.*

More precisely, the A-sequences of an elements of the extended Lagrange, the extended k-Bell, the extended hitting-time, the extended derivative, and the extended checkerboard subgroups of $\mathcal{R}$ are the same as the A-sequences of the corresponding elements of the Lagrange, k-Bell, hitting-time, derivative, and checkerboard subgroups of $\mathcal{R}$, respectively.

Hence, the A^*-sequences of the inverses of an element of the extended Lagrange, the extended k-Bell, the extended hitting-time, the extended derivative, and the extended checkerboard subgroups of $\mathcal{R}$ are the same as the A^*-sequences of inverses of the corresponding element of the A-sequences of the Lagrange, k-Bell, hitting-time, derivative, and checkerboard subgroups of $\mathcal{R}$, respectively.

Proof Let two Riordan matrices $D_1 = (g_1, f_1)$ and $D_2 = (g_2, f_2)$ and their product,

$$D_3 = D_1 D_2 = (g_1 g_2(f_1), f_2(f_1)), \tag{18}$$

have A-sequences, A_1, A_2, and A_3, respectively, where $D_2 \in \mathcal{A}$ and D_1 is an element of the Lagrange, k-Bell, hitting-time, derivative, and checkerboard subgroups of $\mathcal{R}$, respectively. Denote by $A_i(t)$, $i = 1, 2$, and 3, the generating functions of A-sequences of D_i, $i = 1, 2$, and 3, respectively. Then, by noting

$A_2(t) = 1$ and (8), we have

$$A_3(t) = A_2(t)A_1\left(\frac{t}{A_2(t)}\right) = A_1(t),$$

which proves the first half of the proposition. The second half follows the above equation and (10). □

Proposition 5 *Let $Z_3(t)$ be the generating function of the Z-sequence of an element of the extended Lagrange (the extended k-Bell, the extended hitting-time, the extended derivative, or the extended checkerboard subgroups) of $\mathcal{R}$, and let $Z_2(t)$ be the generating function of the Z-sequence of the corresponding element of the Lagrange (the k-Bell, hitting-time, derivative, or checkerboard subgroups) of $\mathcal{R}$. Then,*

$$Z_3(t) = Z_1(t) + (A_1(t) - tZ_1(t))Z_2(t), \tag{19}$$

where $Z_1(t)$ is the Z-sequence of the corresponding element of the Appell subgroup $\mathcal{A}$.

Proof In D_i, $i = 1, 2$, and 3, satisfying Eq. (18), denote by $A_i(t)$, $i = 1, 2$, and 3, the generating functions of A-sequences of D_i, $i = 1, 2$, and 3, respectively, and denote by $Z_i(t)$, $i = 1, 2$, and 3, the generating functions of Z-sequences of D_i, $i = 1, 2$, and 3, respectively. From (9) and noting $A_2(t) = 1$, we have

$$\begin{aligned}Z_3(t) =& \left(1 - \frac{t}{A_2(t)}Z_2(t)\right) Z_1\left(\frac{t}{A_2(t)}\right) + A_1\left(\frac{t}{A_2(t)}\right) Z_2(t)\\ =&(1 - tZ_2(t))Z_1(t) + A_1(t)Z_2(t),\end{aligned}$$

which implies (19). □

Proposition 6 *The vertical recurrence relations of an element of the extended Lagrange, the extended k-Bell, the extended hitting-time, the extended derivative, and the extended checkerboard subgroups of $\mathcal{R}$ are the same as the vertical recurrence relations of the corresponding element of the Lagrange, k-Bell, hitting-time, derivative, and checkerboard subgroups of $\mathcal{R}$, respectively.*

Proof Let (g, t) be an element of $\mathcal{A}$, and let (d, h) be an element of the Lagrange, k-Bell, hitting-time, derivative, and checkerboard subgroups of $\mathcal{R}$. Then, the corresponding element of the extended Lagrange, the extended k-Bell, the extended hitting-time, the extended derivative, and the extended checkerboard subgroups of $\mathcal{R}$ can be represented as

$$(d, h)(g, t) = (dg \circ h, h),$$

which implies that the vertical recurrence relation of (d, h) and the vertical recurrence relation of $(d, h)(g, t)$ are the same due to (12). □

Example 3 For $\left(\frac{1}{1-t}, \frac{t}{1-t}\right) \in \mathcal{B}$ and $\mathcal{H}$ and $\left(\frac{1}{1-t}, t\right) \in \mathcal{A}$, we have

$$\begin{aligned}
&\left(\frac{1}{1-t}, \frac{t}{1-t}\right)\left(\frac{1}{1-t}, t\right) \\
&= \left(\frac{1}{1-t}\frac{1}{1-\frac{t}{1-t}}, \frac{t}{1-t}\right) \\
&= \left(\frac{1}{1-2t}, \frac{t}{1-t}\right).
\end{aligned}$$

The first few entries of the last Riordan array are

$$\left(\frac{1}{1-2t}, \frac{t}{1-t}\right) = \begin{bmatrix} 1 & 0 & 0 & 0 & 0 & 0 & \cdots \\ 2 & 1 & 0 & 0 & 0 & 0 & \\ 4 & 3 & 1 & 0 & 0 & 0 & \\ 8 & 7 & 4 & 1 & 0 & 0 & \\ 16 & 15 & 11 & 5 & 1 & 0 & \\ 32 & 31 & 26 & 16 & 6 & 1 & \\ & & \cdots & & & & \ddots \end{bmatrix}.$$

It is clear that the A-sequences of $\left(\frac{1}{1-t}, \frac{t}{1-t}\right)$ and its extension $\left(\frac{1}{1-2t}, \frac{t}{1-t}\right)$ are the same represented as $A = \{1, 1, 0, \ldots\}$, and the vertical recurrence relations of $\left(\frac{1}{1-t}, \frac{t}{1-t}\right)$ and its extension $\left(\frac{1}{1-2t}, \frac{t}{1-t}\right)$ are the same.

Example 4 For $\left(\frac{1}{(1-t)^2}, \frac{t}{1-t}\right) \in \mathcal{D}$ and $\left(\frac{1}{1-t}, t\right) \in \mathcal{A}$, we have

$$\begin{aligned}
&\left(\frac{1}{(1-t)^2}, \frac{t}{1-t}\right)\left(\frac{1}{1-t}, t\right) \\
&= \left(\frac{1}{(1-t)^2}\frac{1}{1-\frac{t}{1-t}}, \frac{t}{1-t}\right) \\
&= \left(\frac{1}{(1-t)(1-2t)}, \frac{t}{1-t}\right).
\end{aligned}$$

The first few entries of the last Riordan array are

$$\left(\frac{1}{(1-t)(1-2t)}, \frac{t}{1-t}\right) = \begin{bmatrix} 1 & 0 & 0 & 0 & 0 & 0 & \cdots \\ 3 & 1 & 0 & 0 & 0 & 0 & \\ 7 & 4 & 1 & 0 & 0 & 0 & \\ 15 & 11 & 5 & 1 & 0 & 0 & \\ 31 & 26 & 16 & 6 & 1 & 0 & \\ 63 & 57 & 42 & 22 & 7 & 1 & \\ & & \cdots & & & & \ddots \end{bmatrix}.$$

It is clear that the A-sequences of $\left(\frac{1}{(1-t)^2}, \frac{t}{1-t}\right)$ and its extension $\left(\frac{1}{(1-t)(1-2t)}, \frac{t}{1-t}\right)$ are the same represented as $A = \{1, 1, 0, \ldots\}$, and the vertical recurrence relations of $\left(\frac{1}{(1-t)^2}, \frac{t}{1-t}\right)$ and its extension $\left(\frac{1}{(1-t)(1-2t)}, \frac{t}{1-t}\right)$ are the same.

Example 5 We now consider $\left((tC(t))', tC(t)\right) \in \mathcal{D}$ and $\left(\frac{1}{1-t}, t\right) \in \mathcal{A}$. Since $C(t) = 1 + tC(t)^2$, taking derivatives on the both sides of the last equation yields

$$C'(t) = C(t)^2 + 2tC(t)C'(t).$$

By the last equation and $C(t)^2 = (C(t) - 1)/t$,

$$C'(t) = \frac{C - 1}{t(1 - 2tC)}.$$

Thus, we have

$$\begin{aligned} &\left((tC(t))', tC(t)\right)\left(\frac{1}{1-t}, t\right) \\ &= \left(C(t) + tC'(t), tC(t)\right)\left(\frac{1}{1-t}, t\right) \\ &= \left(\frac{1}{1-2tC(t)}, tC(t)\right)\left(\frac{1}{1-t}, t\right) \\ &= \left(\frac{1}{(1-tC(t))(1-2tC(t))}, \frac{t}{1-t}\right). \end{aligned}$$

The first few entries of the last Riordan array are

Anderson and Livingston [1] modified this definition by restricting the vertex set to the nonzero zero-divisors and enriched the subject by proving many graph theoretic and ring theoretic results. For the advancement of the theory of zero divisor graphs, please see [1, 2, 4] and [7].

Uniform hyper-graph is a natural generalization of a simple graph (see page 2 in [5]). In 2007, Ch. Eslahchi and A. M. Rahimi defined k uniform hyper-graphs over a commutative ring as a generalization of zero divisor graphs in [9]. In the process of doing so, they introduced three new concepts in ring theory, k zero divisors, k prime ideals and k integral domains or simply k domains. Along with several elementary observations, their paper also contains the following interesting results;

1. If R is not a k integral domain and has two prime ideals P_1 and P_2 with $P_1 \cap P_2 = \{0\}$, then $H_k(R)$ is a bipartite hyper-graph.
2. If R is a product of n integral domains, then the chromatic number $\chi(H_n(R)) = 2$ and $\chi(H_{n+t}(R)) \leq 2 + t$ for any $t \geq 0$.
3. If $x^2 \neq 0$ for all $x \in Z(R, 3)$, then $H_3(R)$ is connected with a diameter of at most 4.
4. For any finite non-local ring R, the hyper-graph $H_3(R)$ is complete if and only if R is isomorphic to $\mathbb{Z}_2 \times \mathbb{Z}_2 \times \mathbb{Z}_2$.

In the same paper, they have posed the following question:

? Eslahchi-Rahimi Question (Question 1, page 4 in [9])

Does the finiteness of k-zero-divisors in a non-k-integral domain R imply the finiteness of zero-divisors or, equivalently, finiteness of R?:

An affirmative result is known for rings in which (1) there are finitely many but more than k minimal primes and (2) the nilradical is finitely generated (see theorem 3.4 in [11]). This result is generalized in [12] by dropping the second hypothesis.

We end this section by recording some further advancement in the subject. It is shown in [12], that if R is a k domain, then it is an l domain for any integer $l \geq k$. This enable us to define a hyper-graphic constant $HG(R)$, which is the largest integer (if it exists) g such that R is not a g domain. If such g exists, then $H_g(R)$ is called the maximal hyper-graph. The maximal hyper-graph of $\mathbb{Z}_n$ along with their group of automorphisms $Aut(H_g(R))$ are studied in [13]. Note that $HG(\mathbb{Z}_8) = 2$, and $HG(\mathbb{Z}_8 \times \mathbb{Z}_8) = 6$. Thus in general $HG(R \times S) \neq HG(R) + HG(S)$. In Sect. 4, We will present a class of finite rings for which equality holds. Section 2 contains basic definitions and notations. The finiteness theorem is proved in Sect. 3. In Sect. 4, we use some of the results from [6] to classify finite rings with $HG(R) = 2$. Here, we also record some interesting observations on the hyper-graphic constant of local finite rings.

2 Definitions and Notations

All rings considered in this paper are commutative with identity. Most terms from the theory of commutative rings are taken from [3] and from [6]. Graph theoretic notations are adopted from [9]. Our definition of a uniform hyper-graph is the same as introduced in [5]. We now record frequently used terms.

An *ideal* I is a subset of a ring R such that $a + b$ and $r \cdot a$ are in I for all a, b in I and $r \in R$. An ideal P is called a *prime* ideal if given $ab \in P$, either $a \in P$ or $b \in P$. A *maximal* ideal is a proper ideal which is not contained in any other proper ideal. If R has only one maximal ideal $\mathfrak{m}$, then R is called a *local ring* and is denoted by $(R, \mathfrak{m})$.

A *nontrivial idempotent* is an element e of R such that $e \neq 0$, $e \neq 1$ and $e^2 = e$. An element a in R is called a *zero divisor* if there exists a nonzero element b such that $ab = 0$. We will denote the set of zero divisors by $Z(R)$. An element $a \in R$ is called *nilpotent*, if there exists an integer $n > 0$ such that $a^n = 0$. The set of all the nilpotent elements $N(R)$ forms an ideal called the nilradical of R. It is known that $N(R)$ is the intersection of all the prime ideals (see [3], proposition 1.7 and 1.8). Clearly, $N(R) \subset Z(R)$. For any $a \in N(R)$, we define $d_{nil}(a)$ to be the smallest integer $n \geq 0$ such that $a^n = 0$.

The *annihilator* of an element $a \in R$ is an ideal defined as

$$Ann(a) = \{x \in R | a \cdot x = 0\}.$$

Remark Suppose I is an ideal in a ring R. For any $r \in R$, the set $r + I$ is called a *coset* of I in R. Distinct cosets partition R into equivalence classes. If I is finite and has finitely many cosets, then R is finite and $|R| = |I| \cdot$(number of cosets), where $|I|$ is the cardinality of I. When $I = Ann(a)$, then distinct cosets are in one to one ;correspondence with the ideal $a \cdot R$. Hence, in the event when both, $|Ann(a)|$ and $|a \cdot R|$ are finite, R is also finite. We will use this fact in the finiteness theorem.

We recall the following definition from [9].

Definition 1 Let R be a commutative ring and $k \geq 2$ be a fixed integer. A nonzero and non unit element $a_1 \in R$ is called a *k-zero divisor* if there exist non units $a_2, a_3, \ldots, a_k$ in R such that

1. elements in the set $\Lambda = \{a_1, a_2, \ldots, a_k\}$ are pairwise distinct,
2. $\prod \Lambda = \prod_1^k a_i = 0$ and
3. $\prod_{i \neq j} \Lambda = \prod_{i \neq j} a_i \neq 0$ for any j, $1 \leq j \leq k$.

The set of all k-zero divisors is denoted by $Z(R, k)$. Clearly $Z(R, k) \subset Z(R)$. A ring in which $Z(R, k)$ is a null set is called a *k-integral domain* or just a *k-domain*.

Note This definition has a redundancy in demanding a_1 to be a nonzero and non unit. Since $a_1(\prod_2^k a_i) = 0$ and $\prod_2^k a_i \neq 0$, a_1 is not a unit. Further, $\prod_1^{k-1} a_i \neq 0$. Hence, $a_1 \neq 0$. The same is true for all a_i. We drop these requirements in what follows.

We define a k uniform hyper-graph $H_k(R)$ on a ring R as follows. The vertex set is $Z(R, k)$. Sets like $\Lambda = \{a_1, a_2, \ldots, a_k\}$ which appeared in Definition 1 form *k-edges* (edge with k elements) in $H_k(R)$. If a, b belong to a k-edge, then we say that b is *incident* to a. The *degree* of a vertex a is defined as the cardinality of the set of vertices incident to a. A *finite degree edge* is a k-edge which only contains vertices having finite degree. For any k edge Λ, We write

$$V(\Lambda) = \{x \in Z(R, k) \mid x \text{ is incident to some } a \in \Lambda\}.$$

Clearly $\Lambda \subset V(\Lambda)$. Further, it is easy to verify that Λ is a finite degree edge if and only if $V(\Lambda)$ is a finite set.

We present now two examples of uniform hyper-graphs over $\mathbb{Z}_n$. More examples, with detailed descriptions, can be found in [9], [11] and in [12].

Example 1 We consider the ring $\mathbb{Z}_{12} = \{0, 1, \ldots, 11\}$ with operations addition and multiplication modulo 12. The graph $H_3(\mathbb{Z}_{12})$ has the vertex set $Z(\mathbb{Z}_{12}, 3) = \{2, 3, 9, 10\}$. There are two 3-edges, $\{2, 3, 10\}$ and $\{2, 9, 10\}$ which are represented by the two ovals in Fig. 1.

Note that $\mathbb{Z}_{12}$ is a 4- domain.

Example 2 $H_3(\mathbb{Z}_{18})$ has vertex set

$$Z(\mathbb{Z}_{18}, 3) = \{2, 3, 4, 8, 10, 14, 15, 16\}$$

and six edges, $\{2,3,15\}$,$\{4,3,15\}$,$\{8,3,15\}$, $\{10,3,15\}$,$\{14,3,15\}$ and $\{16,3,15\}$, as shown in Fig. 2.

We end this section with a trivial observation, which is used frequently in the next section.

Proposition 1 *Let X be an infinite subset of a ring R. Then $Y = \{x - y \mid x, y \in X\}$ is also infinite.*

Proof Let $x_0 \in X$. Then for any pair $y_1 \neq y_2$, $x_0 - y_1 \neq x_0 - y_2$. Hence $\{x_0 - y \mid y \in X\} \subset Y$ is infinite. This completes the proof. □

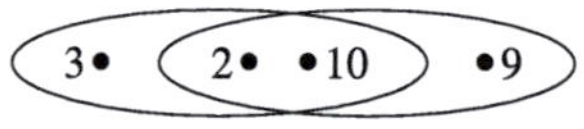

Fig. 1 Three uniform hyper-graph on $\mathbb{Z}_{12}$

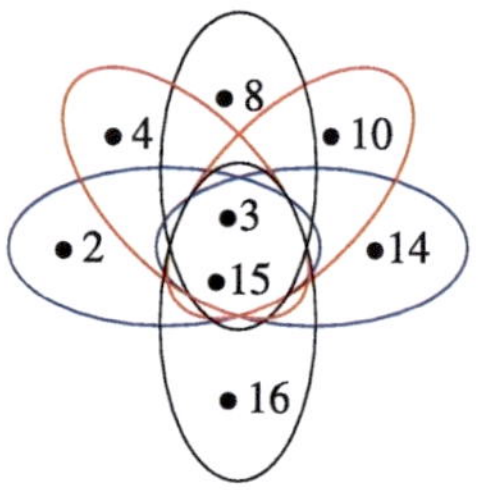

Fig. 2 Three uniform hyper-graph $H_3(\mathbb{Z}_{18})$

3 Finiteness Theorem

In most graph theoretic applications, one expects a graph to be finite or equivalently, the vertex set of a graph to be finite. Therefore it is natural to ask, does the finiteness of $Z(R,k)$ imply the finiteness of R? In general, this is not true. For any infinite integral domain, $Z(R,k) = \varnothing$ for all k. For non integral domains, Ganesan showed in [10], that a ring R which is not an integral domain has finitely many zero divisors if and only if R is finite. We record below a graph theoretic version of this result, which uses essentially the same proof. We recall that the vertex set of the zero divisor graph $\Gamma(R)$ is the set of nonzero zero-divisors $Z(R)\backslash\{0\}$ and a two element set $\{a,\ b\}$ is an edge if $ab = 0$.

Remark A ring which is not an integral domain but is a k domain for any $k > 1$, is isomorphic to $\mathbb{Z}_4$ or $\mathbb{Z}_2[x]/ < x^2 >$ (see [11] for a proof). Both rings are finite, so in the following theorem we will assume that R has at least two nonzero zero divisors a and b such that $ab = 0$, that is, $\Gamma(R) \neq \varnothing$.

Theorem 1 (Finiteness Theorem for Zero Divisor Graphs) *Suppose R is a commutative ring which is not a domain. Then the following statements are equivalent:*

1. *The zero divisor graph $\Gamma(R)$ contains a finite degree edge.*
2. *R is finite.*
3. *R has finitely many zero divisors.*
4. *All vertices of $\Gamma(R)$ have finite degree.*

Proof The implications (2) implies (3), (3) implies (4) and (4) implies (1) are immediate. Hence we only need to prove that (1) implies (2). As we noted in the remark above, we can assume that $\Gamma(R) \neq \varnothing$. Let $\Lambda = \{a,\ b\}$ be a finite degree edge in $\Gamma(R)$. Then $Ann(a) \subset V(\Lambda) \cup \{0, a\}$ which is a finite set. Hence $Ann(a)$ is finite. Similarly, $Ann(b)$ is also finite. But $a \cdot R \subset Ann(b)$, hence $a \cdot R$ is also finite. Now we use the remark made in the beginning of Sect. 2 to derive that R is finite. This completes the proof. □

There are infinite domains, hence non-emptiness of $Z(R,k)$ is a necessary condition.

The following 'finiteness' lemma is proved in [12], which was used to show that every regular element in a ring in which $Z(R,k)$ is nonempty and finite is a unit.

Lemma 1 *Suppose $Z(R,k)$ is finite. Then for any pair $a_1,\ a_2$ of incident vertices, the ideal $I = Ann(a_1) \cap Ann(a_2)$ is finite.*

We will prove a generalization of this lemma, though the proof is similar. In the remainder of this section, we will assume that R is not a k-domain for some $k > 2$ and $H_k(R)$ contains a finite degree edge $\Lambda = \{a_1, a_2, \ldots, a_k\}$.

Lemma 2 *For any pair of integers $r,\ s$, $1 \leq r < s < k$, the ideal*

$$I = Ann(\prod_{1}^{r} a_i) \cap Ann(\prod_{r+1}^{s} a_i)$$

is finite. Further, for any permutation $\sigma \in S_k$, *the ideal* $I_\sigma = Ann(\prod_1^r a_{\sigma(i)}) \cap Ann(\prod_{r+1}^s a_{\sigma(i)})$ *is finite.*

Proof Suppose I is infinite. Since Λ is a finite degree edge, $V(\Lambda)$ is finite. Hence the set

$$B = \{x \in I \mid a_k + x \notin V(\Lambda)\}.$$

is infinite and hence nonempty. For any $x \in B$, let $\Lambda_x = \{a_1, \ldots, a_{k-1}, a_k + x\}$. By our choice of x, all vertices in Λ_x are pairwise distinct and $\prod_1^{k-1} a_i \cdot (a_k + x) = \prod a_i = 0$. To verify condition (3) of Definition 1, first we note that $\prod_1^{k-1} a_i \neq 0$, as Λ is an edge. For any j, $1 \leq j < k$, j is either not in $\{1, 2, \ldots, r\}$ or not in $\{r + 1, \ldots, s\}$. Consequently, $x \in Ann(\prod_{i \neq j}^{k-1} a_i)$. Hence, $\prod_{i \neq j}^{k-1} a_i(a_k + x) = \prod_{i \neq j} a_i \neq 0$. Therefore, Λ_x forms a k-edge and $a_k + x \in V(\Lambda)$, which is a contradiction. Hence I must be a finite set. By reordering Λ, we get the second assertion. □

The above lemma implies that $Ann(a_1) \cap Ann(\prod_2^{k-1})$ is finite. We stretch it further in the following lemma.

Lemma 3 *The ideal*

$$I = Ann(a_1) \cap Ann(\prod_{2}^{k} a_i)$$

is finite. Further, for any $\sigma \in S_k$, $I_\sigma = Ann(a_{\sigma(1)}) \cap Ann(\prod_2^k a_{\sigma(i)})$ *is finite.*

Proof Assume that the ideal I is infinite. By Lemma 2, for all $j > 1$,

$$B_j = Ann(a_1) \cap Ann(\prod_{i \neq j, i \geq 2}^{k} a_i)$$

is finite. Therefore, $B = I \setminus \left(\cup B_j \cup V(\Lambda)\right)$ is infinite and hence nonempty. Now for any $x \in B$, define $\Lambda_x = \{x, a_2, \ldots, a_k\}$. Then the elements in Λ_x are all distinct. Further, $x \cdot (\prod_2^k a_i) = 0$. Now since Λ is a k-edge, $\prod_2^k a_i \neq 0$. Further, for any j, $2 \leq j \leq k$, $x \notin Ann(\prod_{i \neq j, 2}^k a_i)$. Hence, $x \cdot (\prod_{i \neq j, 2}^k a_i) \neq 0$. Therefore, $x \in V(\Lambda)$, which contradicts the choice of x. Hence, I must be a finite ideal. The second assertion has a similar proof. □

In the next lemma we will consider a larger ideal, $Ann(a_1)$ and show that it is finite.

$\subset B \subset I_r$, hence $\{x - x_0 | x \in B_2\} \subset Ann(\prod_1^r a_i) \cap Ann(\prod_{r+1}^{k-1} a_i)$ which s Lemma 2. Therefore, I must be finite. This completes the proof. □

an prove the finiteness theorem.

2 *Suppose $k \geq 2$ and R is a ring with nonempty $Z(R, k)$. Then the are equivalent;*

graph $H_k(R)$ contains a finite degree edge.
ite.
) is finite.
vertices of $H_k(R)$ have a finite degree.

plications (2) implies (3), (3) implies (4) and (4) implies (1) are evident. only need to prove (1) implies (2). Case $k = 2$ is covered in Theorem 1, can assume that $k > 2$. Let $\Lambda = \{a_1, a_2, \ldots, a_k\}$ be a finite degree k-Lemma 4, $Ann(a_1)$ is finite. Now since $a_1 \in Ann(\prod_2^k a_i)$, the principal $\subset Ann(\prod_2^k a_i)$. By Lemma 4, $Ann(\prod_2^k a_i)$ is finite, hence $a_1 R$ is also ce, both $Ann(a_1)$ and $a_1 \cdot R$ are finite, the remark given in the beginning completes the proof. □

ite Rings and Hyper-Graphic Constants

ection, we will consider finite commutative rings. The classification of gs is a computationally complex problem. Additive structure can be easily using the fundamental theorem of abelian groups, but it is possible for two orphic rings to have isomorphic additive structures. For example, $\mathbb{Z}_2 \times \mathbb{Z}_2$ $]/ < x^2 >$ have isomorphic additive structures but their multiplicative is not. First ring has a non-trivial idempotent.

eral, we can define a ring homomorphism ϕ from $\mathbb{Z}$ to R by sending 1 to ity of R and extending it additively. Image $\phi(\mathbb{Z})$ is a sub-ring of R, which the *prime sub-ring* of R. When $Char(R) = 0$, the prime sub-ring is $\mathbb{Z}$. ite ring R, the prime sub-ring is $\mathbb{Z}_n$ for some n. If n is composite, then R a nontrivial idempotent, which splits R in to two factor rings. That is, if e rivial idempotent, then $R \simeq R_1 \times R_2$, where $R_1 \simeq eR$ and $R_2 \simeq (1 - e)R$. ely we can find finitely many rings $R_1, R_2, \ldots, R_n$, such that

$$R \simeq R_1 \times R_2 \times \ldots \times R_n$$

each i, R_i does not contain any nontrivial idempotent. In other words, R_i rings. We quote now a general structure theorem of finite rings, see [8] y 2.16) or [6], (theorem 3.1.4) for the complete proof.

n 3 *Any finite ring is isomorphic to a product of finite local rings.*

Lemma 4 *For any i, $1 \leq i \leq k$, $Ann(a_i)$ is a fin*

Proof By reordering if needed, we can assume
an infinite ideal. Since, by Lemma 3, the set An
complementary set $I = Ann(a_1) \backslash Ann(\prod_2^k a_i)$ is
given in lemma 2, we can construct an infinite s
edge $\Lambda_x = \{a_1, a_2, \ldots, a_{k-1}, a_k + x\}$ satisfies fi
Since Λ_x is not an edge, $\prod_{i \neq j}^{k-1} a_i (a_k + x) = 0$.

But Λ is a k-edge, hence $\prod_1^{k-1} a_i \neq 0$. Furthe

$$a_1 \cdot \prod_{i \neq j, 2}^{k-1} a_i (a_k + x) = \left(\prod_{i \neq j, 1}^{k} a_i \right) + \left(\prod_{i \neq j, 2}^{k-1} \right.$$

So j must be 1 and $\prod_2^{k-1} a_i (a_k + x) = 0$ or $a_k +$
any $y \in B$, $a_k + y$ is also in $Ann(\prod_2^{k-1} a_i)$. There

$$x - y = (a_k + x) - (a_k + y) \in$$

Since $B \subset Ann(a_1)$, $x - y \in Ann(a_1) \cap Ann(\prod_2^{k-1}$
is infinite, we get a contradiction to Lemma 2. Thu

Next we will show that $Ann(\prod_i a_i)$ is finite, whi
theorem.

Lemma 5 *For any r, $1 \leq r \leq k - 1$, $I_r = Ann($
$\sigma \in S_k$, $I_{\sigma,r} = Ann(\prod_1^r a_{\sigma(i)})$ is finite.*

Proof We will use induction on r. Case $r = 1$ i
Assume that $r > 1$ and that the result is true for
permutation). Suppose $I_r = Ann(\prod_1^r a_i)$ is an infi
j, $Ann(\prod_{i \neq j}^r a_i)$ is finite. Hence, $I = I_r \backslash \cup_j Ann$
again we use arguments presented in Lemmas 2 and
B of I such that for all $x \in B$, $\prod_{i \neq j}^{k-1} a_i (a_k + x) =$

Now if $r < j < k$, then since $x \in Ann(\prod_1^t a_i)$, $\prod$
Hence $1 \leq j \leq r$. But B is infinite. Therefore, we ca
and a fixed index $j_0 \leq r$ such that for all $x \in B_1$, $\Big(\prod$
0.

Set $w(x) = \prod_{r+1}^{k-1} a_i (a_k + x)$. Then $w(x) \in A$
hypothesis, $Ann(\prod_{i \neq j_0}^r a_i)$ is finite. Therefore, the s
Since B_1 is infinite, there exists $x_0 \in B_1$ such tha
many $x \in B_1$. Let $B_2 = \{x \in B_1 \mid w(x) = w(x$
each $x \in B_2$, $\prod_{r+1}^{k-1} a_i (a_k + x) = \prod_{r+1}^{k-1} a_i (a_k + x_0)$

$B_2 \subset B$
contradi

Now we

Theore
followin

1. *Hype*
2. *R is*
3. *$Z(R,$*
4. *All th*

Proof I
Thus, w
hence v
edge. B
ideal a_1
finite. S
of Sect.

4 Fir

In this
finite ri
classifie
non iso
and $\mathbb{Z}_2$
structur

In g
the ider
is calle
For a fi
contain
is a non
Inducti

and for
are loc
(coroll

Theor

$B_2 \subset B_1 \subset B \subset I_r$, hence $\{x - x_0 | x \in B_2\} \subset Ann(\prod_1^r a_i) \cap Ann(\prod_{r+1}^{k-1} a_i)$ which contradicts Lemma 2. Therefore, I must be finite. This completes the proof. □

Now we can prove the finiteness theorem.

Theorem 2 *Suppose $k \geq 2$ and R is a ring with nonempty $Z(R, k)$. Then the following are equivalent;*

1. *Hyper-graph $H_k(R)$ contains a finite degree edge.*
2. *R is finite.*
3. *$Z(R, k)$ is finite.*
4. *All the vertices of $H_k(R)$ have a finite degree.*

Proof Implications (2) implies (3), (3) implies (4) and (4) implies (1) are evident. Thus, we only need to prove (1) implies (2). Case $k = 2$ is covered in Theorem 1, hence we can assume that $k > 2$. Let $\Lambda = \{a_1, a_2, \ldots, a_k\}$ be a finite degree k-edge. By Lemma 4, $Ann(a_1)$ is finite. Now since $a_1 \in Ann(\prod_2^k a_i)$, the principal ideal $a_1 R \subset Ann(\prod_2^k a_i)$. By Lemma 4, $Ann(\prod_2^k a_i)$ is finite, hence $a_1 R$ is also finite. Since, both $Ann(a_1)$ and $a_1 \cdot R$ are finite, the remark given in the beginning of Sect. 2 completes the proof. □

4 Finite Rings and Hyper-Graphic Constants

In this section, we will consider finite commutative rings. The classification of finite rings is a computationally complex problem. Additive structure can be easily classified using the fundamental theorem of abelian groups, but it is possible for two non isomorphic rings to have isomorphic additive structures. For example, $\mathbb{Z}_2 \times \mathbb{Z}_2$ and $\mathbb{Z}_2[x]/ < x^2 >$ have isomorphic additive structures but their multiplicative structure is not. First ring has a non-trivial idempotent.

In general, we can define a ring homomorphism ϕ from $\mathbb{Z}$ to R by sending 1 to the identity of R and extending it additively. Image $\phi(\mathbb{Z})$ is a sub-ring of R, which is called the *prime sub-ring* of R. When $Char(R) = 0$, the prime sub-ring is $\mathbb{Z}$. For a finite ring R, the prime sub-ring is $\mathbb{Z}_n$ for some n. If n is composite, then R contains a nontrivial idempotent, which splits R in to two factor rings. That is, if e is a nontrivial idempotent, then $R \simeq R_1 \times R_2$, where $R_1 \simeq eR$ and $R_2 \simeq (1 - e)R$. Inductively we can find finitely many rings $R_1, R_2, \ldots, R_n$, such that

$$R \simeq R_1 \times R_2 \times \ldots \times R_n$$

and for each i, R_i does not contain any nontrivial idempotent. In other words, R_i are local rings. We quote now a general structure theorem of finite rings, see [8] (corollary 2.16) or [6], (theorem 3.1.4) for the complete proof.

Theorem 3 *Any finite ring is isomorphic to a product of finite local rings.*

Lemma 4 *For any i, $1 \le i \le k$, $Ann(a_i)$ is a finite ideal.*

Proof By reordering if needed, we can assume that $i = 1$. Suppose $Ann(a_1)$ is an infinite ideal. Since, by Lemma 3, the set $Ann(a_1) \cap Ann(\prod_2^k a_i)$ is finite, the complementary set $I = Ann(a_1) \backslash Ann(\prod_2^k a_i)$ is infinite. Now using the arguments given in lemma 2, we can construct an infinite set B such that for all $x \in B$, the edge $\Lambda_x = \{a_1, a_2, \ldots, a_{k-1}, a_k + x\}$ satisfies first two conditions of Definition 1. Since Λ_x is not an edge, $\prod_{i \ne j}^{k-1} a_i(a_k + x) = 0$.

But Λ is a k-edge, hence $\prod_1^{k-1} a_i \ne 0$. Further,

$$a_1 \cdot \prod_{i \ne j, 2}^{k-1} a_i(a_k + x) = \left(\prod_{i \ne j, 1}^{k} a_i\right) + \left(\prod_{i \ne j, 2}^{k-1} a_i\right) \cdot a_1 \cdot x = \prod_{i \ne j, 1}^{k} a_i \ne 0.$$

So j must be 1 and $\prod_2^{k-1} a_i(a_k + x) = 0$ or $a_k + x \in Ann(\prod_2^{k-1} a_i)$. Similarly, for any $y \in B$, $a_k + y$ is also in $Ann(\prod_2^{k-1} a_i)$. Therefore, the difference

$$x - y = (a_k + x) - (a_k + y) \in Ann(\prod_2^{k-1} a_i).$$

Since $B \subset Ann(a_1)$, $x - y \in Ann(a_1) \cap Ann(\prod_2^{k-1} a_i)$. As the set $\{x - y \mid x, y \in B\}$ is infinite, we get a contradiction to Lemma 2. Thus I must be a finite ideal. □

Next we will show that $Ann(\prod_i a_i)$ is finite, which will lead us to the finiteness theorem.

Lemma 5 *For any r, $1 \le r \le k - 1$, $I_r = Ann(\prod_1^r a_i)$ is finite. Further, for any $\sigma \in S_k$, $I_{\sigma,r} = Ann(\prod_1^r a_{\sigma(i)})$ is finite.*

Proof We will use induction on r. Case $r = 1$ is already covered in Lemma 4. Assume that $r > 1$ and that the result is true for all integers less than r (up to a permutation). Suppose $I_r = Ann(\prod_1^r a_i)$ is an infinite ideal. By induction, for any j, $Ann(\prod_{i \ne j}^r a_i)$ is finite. Hence, $I = I_r \backslash \cup_j Ann(\prod_{i \ne j}^r a_i)$ is also infinite. Once again we use arguments presented in Lemmas 2 and 4 to construct an infinite subset B of I such that for all $x \in B$, $\prod_{i \ne j}^{k-1} a_i(a_k + x) = 0$ for some j, $1 \le j < k$.

Now if $r < j < k$, then since $x \in Ann(\prod_1^t a_i)$, $\prod_{i \ne j}^{k-1} a_i(a_k + x) = \prod_{i \ne j}^{k} a_i \ne 0$. Hence $1 \le j \le r$. But B is infinite. Therefore, we can find an infinite subset B_1 of B and a fixed index $j_0 \le r$ such that for all $x \in B_1$, $\left(\prod_{i \ne j_0}^r a_i\right) \cdot \left(\prod_{r+1}^{k-1} a_i(a_k + x)\right) = 0$.

Set $w(x) = \prod_{r+1}^{k-1} a_i(a_k + x)$. Then $w(x) \in Ann(\prod_{i \ne j_0}^r a_i)$. By the inductive hypothesis, $Ann(\prod_{i \ne j_0}^r a_i)$ is finite. Therefore, the set $\{w(x) | x \in B_1\}$ is also finite. Since B_1 is infinite, there exists $x_0 \in B_1$ such that, $w(x) = w(x_0)$ for infinitely many $x \in B_1$. Let $B_2 = \{x \in B_1 \mid w(x) = w(x_0)\}$. Then B_2 is infinite and for each $x \in B_2$, $\prod_{r+1}^{k-1} a_i(a_k + x) = \prod_{r+1}^{k-1} a_i(a_k + x_0)$ or $\prod_{r+1}^{k-1} a_i(x - x_0) = 0$. Now

We will denote a local ring by $(R, \mathfrak{m})$, where $\mathfrak{m}$ is the maximal ideal of R. If $char(R) = p^{\alpha}$, then $\mathbb{Z}_{p^{\alpha}}$ is a sub-ring of R. Furthermore, R contains a *coefficient ring* $T = GR(p^{\alpha}, r)$, which is an extension of $\mathbb{Z}_{p^{\alpha}}$ by a primitive root of unity. The ring $GR(p^{\alpha}, r)$ is called a Galois ring. We will state this result for future reference, see [8] (remark after theorem 7.8) or [6], (theorems 6.3.1) for the detailed discussion.

Theorem 4 *Let* $(R, \mathfrak{m})$ *be a finite local ring with* $Char(R) = p^{\alpha}$ *for some prime* p*. Then* R *contains a coefficient ring* $T = \mathbb{Z}_{p^{\alpha}}[\xi]$ *such that*

$$R \sim T[x_1, \ldots, x_r]/I$$

for some positive integer r *and an ideal* I *in* $T[x_1, \ldots, x_r]$.

Thus R is a quotient of a multivariate polynomial ring over the coefficient ring T. The maximal ideal $\mathfrak{m}$ is generated by the image $\bar{x_i}$ and the prime p.

Recall that the hyper-graphic constant $HG(R)$ of a ring R is an integer g such that $H_g(R) \neq \emptyset$ but $H_l(R) = \emptyset$ for all $l > g$. If no such g exist, then we say that $HG(R) = \infty$. We also define $HG(R) = 0$ for domains, and $HG(R) = 1$ for rings which are not domains, but are 2-domains.

In general, the hyper-graphic constant is not additive. However, we will show that for a certain class of finite rings, it is additive. First we recall that the nilradical $N(R)$ is an ideal consisting of elements $a \in R$ such that $a^n = 0$ for some integer $n \geq 1$. Nil degree of an element $a \in N(R)$, denoted by $d_{nil}(a)$ is the smallest positive integer n such that $a^n = 0$. Finally, for a prime ideal P, R_P is a localization of R at P (see chapter 3 in [3] for the detailed description of R_P). In the context of finite rings, R_P is one of the factor rings present in the decomposition of R (see the discussion leading to Theorem 3). In fact, the set $Spec R$ of prime ideals and the set of factor rings are in one to one correspondence. We start with a lemma.

Lemma 6 *Let* R *be a commutative ring (need not be finite) and* $a \in N(R)$*. If* $d_{nil}(a) = n$ *for some* $n \geq 4$*, then* $HG(R) \geq n$.

Proof By definition, $a^{n-1} \neq 0$. Define $x_i = \sum_{j=1}^{i} a^j$, for $1 \leq i \leq n-1$ and $x_n = a + a^3$. We will verify three conditions of the Definition 1 for $\Lambda = \{x_1, x_2, \ldots, x_n\}$.

To show that elements of Λ are pairwise distinct, let us assume that $x_i = x_j$ for some $i, j, 1 \leq i < j < n$. Then

$$a + a^2 + \ldots + a^i = a + a^2 + \ldots + a^j.$$

Cancellation gives $0 = a^{i+1} \cdot unit$. Since $i < j < n$, $i + 1 \leq n - 1$, hence $a^{n-1} = 0$ which is not the case. If $j = n$, then we get $0 = a^2 + a^4 + \ldots + a^i$. Again we get $a^2 = 0$ which is a contradiction. Hence, the elements of Λ are pairwise distinct. For the second and third conditions, note that $\prod x_i = a^n \cdot unit = 0$ and $\prod_{i \neq j} x_i = a^{n-1} \cdot unit \neq 0$. Thus Λ is an n edge and $H_n(R)$ is nonempty. Therefore, $HG(R) \geq n$. □

When $HG(R) = 2$, Lemma 6 may not hold. For example, $HG(\mathbb{Z}_8) = 2$, but $2^2 \neq 0$. A direct application of Lemma 6 to a local ring gives us the following result.

Lemma 7 *Suppose* $(R, \mathfrak{m})$ *is a finite local ring with* $HG(R) = n$. *Then for any* $a \in \mathfrak{m}$, *(a)* $a^n = 0$ *if* $n \geq 3$ *and (b)* $a^3 = 0$ *if* $n = 1$ *or* 2.

Proof First observe that elements of $\mathfrak{m}$ are nilpotent. Therefore, for any $a \in \mathfrak{m}$, $a^r = 0$ for some r. If $n \geq 4$, then $r \leq n$ by Lemma 6. Suppose now $n = 2$ or 3. If $r > 3$, then by Lemma 6, $HG(R) \geq r > 3$, which is a contradiction. □

Proposition 2 *Suppose* $(R, \mathfrak{m})$ *is a local ring with* $HG(R) = n$ *for some* $n \geq 3$. *Then* $\mathfrak{m}^n = 0$.

Proof For a given r, let $X(r)$ denote the set of all non-increasing ordered sequences of non-negative integers whose sum is at least n. That is,

$$X(r) = \{(\lambda_1, \ldots, \lambda_r) | \lambda_i \in \mathbb{Z},\ \lambda_i \geq 0,\ \sum \lambda_i \geq n,\ \lambda_i \geq \lambda_j \text{ for } i < j\}.$$

For any $\lambda = (\lambda_1, \ldots, \lambda_r)$ and $\mu = (\mu_1, \ldots, \mu_r)$ in $X(r)$, we say that $\lambda < \mu$ if

(1) $\sum \lambda_i < \sum \mu_i$ or
(2) $\sum \lambda_i = \sum \mu_i$, $\lambda_i = \mu_i$ for $1 \leq i < k$ and $\lambda_k < \mu_k$ for some k.

Let $a_1, \ldots, a_r$ be distinct elements in $\mathfrak{m}$ and $\lambda_i \geq 1$ are integers such that $\sum \lambda_i \geq n$. It suffices to prove that $\prod a_i^{\lambda_i} = 0$. By rearranging a_i, if required, we can assume that $(\lambda_1, \ldots, \lambda_r) \in X(r)$. If $\sum \lambda_i \geq r \cdot n$, then $\lambda_1 \geq n$. By Lemma 6, $a_1^n = 0$, hence $\prod a_i^{\lambda_i} = 0$.

Suppose $\alpha = (\alpha_1, \ldots, \alpha_r) \in X(r)$ is the largest element such that $x = \prod a_i^{\alpha_i} \neq 0$. Define
$A_j = \{a_j, a_j + a_j^2, \ldots, a_j + a_j^2 + \ldots + a_j^{\alpha_j}\}$, then

$$\prod A_j = a_j^{\alpha_j}(1 + a_j)(1 + a_j + a_j^2) \cdot (1 + a_j + \ldots a_j^{\alpha_j - 1}) = a_j^{\alpha_j} \cdot unit$$

hence, $\prod_{j=1}^r (\prod A_j) = x \cdot unit$. Further, $\sum |A_j| = \sum \alpha_i \geq n$. Consider

$$\Lambda_0 = (\cup_1^r A_j) \cup \{a_1 + a_r\}.$$

Observe that $\prod \Lambda_0 = x \cdot unit \cdot (a_1 + a_r) = 0$ by the maximality of α.

Now we will verify that all the elements in Λ_0 are distinct. Suppose $v, w \in \Lambda_0$ are such that $v = w$. Suppose $v \in A_i$ and $w \in A_j$.If $i = j$, then clearly $v \neq w$. Suppose $i < j$. Then $v = a_i \cdot u_1$ and $w = a_j \cdot u_2$ for some units. Substituting $a_i = a_j \cdot u_3$ in x yields

$$x = \prod_{l \neq i} a_l^{\alpha_l} \cdot a_j^{\alpha_j} \cdot u = 0$$

by the maximality of α. Since $x \neq 0$, we must have $v \neq w$. Similar arguments can show that all the elements in Λ_0 are pairwise distinct.

Since $|\Lambda_0| > n$, the last condition of the Definition 1 must fail. That is $\prod_{w \in \Lambda_0,\, w \neq v} w = \prod \Lambda_0/v = 0$ for some $v \in \Lambda_0$.

Now, for any $v \in A_j$,

$$\prod \Lambda_0/v = \begin{cases} x(a_r/a_j),\ if\ 1 < j \leq r \\ x + x(a_r/a_1),\ j = 1. \end{cases}$$

Thus, $x(a_r/a_j) = 0$ for some $1 < j \leq r$, or $x + x(a_r/a_1) = 0$. If $x(a_r/a_j) \neq 0$ for all j, $1 < j \leq r$, then $x + x(a_r/a_1) = 0$ which shows that

$x = -x(a_r/a_1) \neq 0$.

We claim that $x(a_r/a_j) \neq 0$ for all j, $1 \leq j < r$. By the above observation, we only need to prove the claim for $j > 1$.

Assume that $x(a_r/a_j) = 0$ for some j. Let k_1 be the largest integer such that $x(a_r/a_{k_1}) = 0$.

Consider $\Lambda_1 = (\cup_1^r A_j) \cup \{a_1 + a_{k_1} + a_r\}$.

Then $\prod \Lambda_1 = x \cdot unit \cdot (a_1 + a_{k_1} + a_r) = 0$ by the maximality of α. Further, using the arguments similar to one we used for Λ_0, we can show that all the elements in Λ_1 are pairwise distinct.

Since $|\Lambda_1| > n$, the last condition of the Definition 1 must fail.

Now, for any $v \in A_j$,

$$\prod \Lambda_1/v = \begin{cases} x(a_r/a_j) \neq 0,\ if\ k_1 < j \leq r \\ x + x(a_r/a_j) \neq 0,\ if\ j = k_1 \\ x(a_{k_1}/a_j) + x(a_r/a_j),\ if\ 1 < j < k_1 \\ x + x(a_{k_1}/a_1) + x(a_r/a_1),\ j = 1. \end{cases}$$

Hence $x(a_{k_1}/a_j) + x(a_r/a_j) = 0$ for some j, $1 < j < k_1$ or $x + x(a_{k_1}/a_1) + x(a_r/a_1) = 0$. Suppose k_2 is the largest integer such that $x(a_{k_1}/a_{k_2}) + x(a_r/a_{k_2}) = 0$. By the construction, we have $k_2 < k_1$.

Consider now,

$$\Lambda_2 = (\cup A_j) \cup \{a_1 + a_{k_2} + a_{k_1} + a_r\}$$

Then $\prod \Lambda_2 = 0$ by the maximality of α. Further, elements of Λ_2 are pairwise distinct. Therefore the third condition must fail.

Note that, for any $v \in A_j$,

Consider $\Lambda = \{x, x(1+x), x(1+y)\}$. Then $\prod \Lambda = x^3 \cdot (1+x)(1+y) = 0$. Further, the product of any two elements of Λ is $x^2 \cdot unit \neq 0$. Clearly, $x \neq x + x^2$. Therefore, either $x = x + xy$ or $x + x^2 = x + xy$. That is either $xy = 0$ or $x^2 = xy$. Suppose $xy = 0$. Consider $\Lambda_1 = \{x, x(1+x), x+y\}$. Then, $\prod \Lambda_1 = (x^3 + x^2 y)(1+x) = 0$. Also, since $y \notin R_0$, all three elements are distinct. To verify the third property of the Definition 1, observe that

$$x \cdot x(1+x) = x^2 \cdot unit \neq 0$$

$$x \cdot (x+y) = x^2 \neq 0$$

$$x(1+x) \cdot (x+y) = x^2 \cdot unit \neq 0$$

Thus, Λ_1 is a 3-edge, which is a contradiction. Hence, $xy \neq 0$. Suppose that $x^2 = xy$.

Consider $\Lambda_2 = \{x, x(1+x), y\}$. Then, $\prod \Lambda_2 = x^2 y(1+x) = x^3(1+x) = 0$. Moreover, since $y \notin R_0$, all three elements are pairwise distinct. To verify the third property of Definition 1, observe that

$$x \cdot x(1+x) = x^2 \cdot unit \neq 0$$

$$x \cdot y = x^2 \neq 0$$

$$x(1+x) \cdot y = x^2 \cdot unit \neq 0$$

Thus λ_2 is a 3-edge, which is a contradiction. Hence no such y exist and $R = R_0$. □

Proposition 5 *Suppose $(R, \mathfrak{m})$ is a finite local ring with $HG(R) = 2$. If R contains a zero divisor x such that $x^2 \neq 0$, then R is isomorphic to one of the following rings:*

1. $\mathbb{Z}_8$,
2. $\mathbb{Z}_4[x]/ < x^2 + 2, 2x >$ *or*
3. $\mathbb{Z}_2[x]/ < x^3 >$.

Proof It is easy to verify that the rings listed above are finite local rings with $HG(R) = 2$. Now to prove the proposition, let $(R, \mathfrak{m})$ be a finite local ring with $HG(R) = 2$ and $x \in \mathfrak{m}$ be such that $x^2 \neq 0$. We first prove that $2 \in \mathfrak{m}$. To show this, let us assume that $2x \neq 0$. Now consider $\Lambda = \{x,\ x + x^2,\ 2x\}$. Clearly, elements of Λ are pairwise distinct and $\prod \Lambda = 0$. Since $HG(R) = 2$, product of two elements of Λ must be zero. But $x \cdot (x + x^2) = x^2 \cdot unit \neq 0$. Therefore, $x \cdot 2x = 0$ or $x(1+x) \cdot 2x = 0$. But $1 + x$ is a unit. Hence, $2x^2 = 0$, which proves that $2 \in \mathfrak{m}$.

If $char(R) = 2^3$, then the prime ring $\mathbb{Z}_8 \subset R$. By Lemma 8, $R \simeq \mathbb{Z}_8$. Suppose $char(R) = 2^2$. Then $\mathbb{Z}_4 \subset R$ is a prime sub-ring. Since $x^2 \neq 0$, $x \notin \mathbb{Z}_4$. By Lemma 8, $R \simeq \mathbb{Z}_4 < x >$, where $\mathbb{Z}_4 < x >$ is the sub-ring of R generated by x over $\mathbb{Z}_4$. Next, we will show that $2x = 0$. Consider $\Lambda = \{x,\ 3x,\ x(1+x)\}$. Then

$\prod \Lambda = 0$ and $\prod_{i \neq j} \Lambda = x^2 \cdot unit \neq 0$. Hence either $x = 3x$ or $3x = x(1+x)$. Suppose $3x = x(1+x)$ or $2x = x^2$. Consider $\Lambda_0 = \{2, x, x(1+x)\}$. Then

$$\prod \Lambda_0 = 2x^2(1+x) = x^3(1+x) = 0.$$

Clearly, $2x = x^2 \neq 0$, $x(x+x^2) \neq 0$ and since 3 is a unit, $2x + 2x^2 = x^2 + 2x^2 = 3x^2 \neq 0$. Further, $2 \neq x$ and $x \neq x + x^2$. If $2 = x + x^2$, then $2 = 3x$ implying $x \in \mathbb{Z}_4$, which is not the case. Thus Λ_0 is a 3-edge which contradicts the hypothesis. Thus, our assumption $x^2 = 2x$ is not true and $2x$ must be zero. Finally, consider $\Lambda_1 = \{x, 2+x, x(1+x)\}$. Then $\prod \Lambda_1 = 0$ and $\prod_{i \neq j} \Lambda_1 = x^2 \cdot unit \neq 0$. Now $x \neq 2 + x$ and $x \neq x + x^2$. But Λ_1 cannot be a 3-edge. Hence, $2 + x = x + x^2$ or $2 = x^2$. This shows that $R \simeq \mathbb{Z}_4[x]/ < x^2 + 2, 2x >$.

To complete the proof, let us assume that $char(R) = 2$. Then $\mathbb{Z}_2 \subset R$ which is a field. Let $x \in R$ such that $d_{nil}(x) = 3$. Then $R \simeq \mathbb{Z}_2[x]/ < x^3 >$. □

In Proposition 2, we assume that $n \geq 3$ to prove that $\mathfrak{m}^n = 0$. The theorem is true for $n = 2$ with an additional condition.

Proposition 6 *Suppose* $(R, \mathfrak{m})$ *is a local ring with* $HG(R) = 2$. *If for all* $a \in \mathfrak{m}$, $a^2 = 0$, *then* $\mathfrak{m}^2 = 0$.

Proof Suppose $a, b \in \mathfrak{m}$ are such that $ab \neq 0$. Since $a^2 = 0$, $a \neq \pm b$. But then $\{a, b, a+b\}$ forms a 3 edge, which contradicts $HG(R) = 2$. □

In proposition 5, we considered rings which contain a zero divisor x such that $x^2 \neq 0$. If R does not contain such x, then for all $a \in \mathfrak{m}$, $a^2 = 0$. Applying proposition 6, we get $\mathfrak{m}^2 = 0$. For such rings we have the following proposition.

Proposition 7 *Suppose* $(R, \mathfrak{m})$ *is a finite local ring with* $HG(R) = 2$. *If for all* $x \in \mathfrak{m}$, $x^2 = 0$, *then* R *is isomorphic to one of the following rings:*

1. $F_{p^\alpha}[x_1, x_2, \ldots, x_r]/ < x_i \cdot x_j >$ *where,* $1 \leq i, j \leq r$, p *is a prime,* $r \geq 1$ *and* F_{p^α} *is a finite field other than* $\mathbb{Z}_2$,
2. $\mathbb{Z}_2[x_1, x_2, \ldots, x_r]/ < x_i \cdot x_j >$ *where,* $1 \leq i, j \leq r$ *and* $r \geq 2$,
3. $\mathbb{Z}_{p^2}[\xi][x_1, x_2, \ldots, x_r]/ < x_i \cdot x_j, p \cdot x_j >$ *where,* $r \geq 0$, $1 \leq i, j \leq n$ *and* $\mathbb{Z}_{p^2}$ *is a coefficient ring other than* $\mathbb{Z}_4$, *or*
4. $\mathbb{Z}_4[x_1, x_2, \ldots, x_r]/ < x_i \cdot x_j, 2 \cdot x_i >$ *where,* $r \geq 1$, $1 \leq i, j \leq r$.

Proof All rings listed above are local with $\mathfrak{m}^2 = 0$. Hence, $HG(R) \leq 2$. It is easy to produce a 2-edge for each ring, therefore, $HG(R) = 2$. Now, we will show that any ring $(R, \mathfrak{m})$ satisfying the hypothesis must be isomorphic to one of these rings. Since $\mathfrak{m}^2 = 0$, the characteristic of R is p or p^2 for some prime p.

When $Char(R) = p, \mathbb{Z}_p \subset R$. By Theorem 4, R contains a coefficient field F_{p^α} which is isomorphic to $R/\mathfrak{m}$. Let $x_1, x_2, \ldots, x_r$ be a minimal generating set of $\mathfrak{m}$ (that is, the image of these elements form a basis of $\mathfrak{m}$ as a vector space over $R/\mathfrak{m}$). Then $R \simeq F_{p^\alpha}[x_1, \ldots, x_r]/I$ for some ideal I. Since $\mathfrak{m}^2 = 0$, $x_i \cdot x_j \in I$ for all

i, j. But $x_1, x_2, \ldots, x_r$ are linearly independent modulo $\mathfrak{m}$. Hence, $I =< x_i \cdot x_j >$. Note that $HG(\mathbb{Z}_2[x]/ < x^2 >) = 1$. Therefore, when $F_{p^\alpha} = \mathbb{Z}_2, r \geq 2$.

Suppose $char(R) = p^2$. Then R contains a coefficient ring $T = \mathbb{Z}_{p^2}[\xi]$ and $R \simeq T[x_1, x_2, \ldots, x_r]/I$ for some ideal I, where ξ is a primitive root. As before, we can show that $I =< x_i x_j, px_i >$. Note that, when $T = \mathbb{Z}_4, n \geq 1$. □

To complete the classification, we will now turn our attention to non-local rings.

Proposition 8 *Suppose R is a finite ring with $HG(R) = 2$. If R is not local then, R is isomorphic to one of the following rings:*

1. $F_{p^\alpha} \times F_{q^\beta}$,
2. $\mathbb{Z}_2 \times \mathbb{Z}_4$ *or*
3. $\mathbb{Z}_2 \times \mathbb{Z}_2[x]/ < x^2 >$,

where p, q are primes and F_{p^α} is a finite field.

Proof Suppose that the ring R is isomorphic to a product of n domains. Then by the example 2.3 (4) in [9], $HG(R) = n$. Thus, rings in (1) have $HG(R) = 2$. It is easy to verify that the rings in (2) and (3) also have $HG(R) = 2$. Now let us assume that R is a non-local ring with $HG(R) = 2$. By the Theorem 3, R is a product of local rings. Since $HG(R) = 2$, we have $R \simeq R_1 \times R_2$ for some local rings R_1 and R_2. We claim that either R_1 or R_2 must be a field.

Suppose neither are fields. Then there exist nonzero elements x_1, x_2 in R_1 and y_1, y_2 in R_2 such that $x_1 \cdot x_2 = 0$ in R_1 and $y_1 \cdot y_2 = 0$ in R_2.

Consider $\Lambda = \{(x_1, 1), (x_2, y_1), (1, y_2)\}$. Clearly, Λ satisfies all the three conditions of Definition 1. But R is a three domain, which is a contradiction. Therefore, either R_1 or R_2 must be a field.

Suppose $R \simeq F_{p^\alpha} \times R_2$, where R_2 is not a field. Then, R_2 contains x_1, x_2 such that $x_1 \cdot x_2 = 0$. Consider $\Lambda_1 = \{(1, x_1), (1.x_2), (0, 1)\}$. Then Λ satisfies conditions (2) and (3) of the Definition 1. Since R is a 3-domain, we must have $x_1 = x_2$. But this shows that $HG(R_2) = 1$. By proposition 2.1 in [12], R_2 is isomorphic to $\mathbb{Z}_4$ or $\mathbb{Z}_2[x]/ < x^2 >$. Thus, R is isomorphic to either $F_{p^\alpha} \times \mathbb{Z}_4$ or $F_{p^\alpha} \times \mathbb{Z}_2[x]/ < x^2 >$.

Now suppose $R \simeq F_{p^\alpha} \times \mathbb{Z}_4$. If F_{p^α} contains a unit u, $u \neq 1$, then $\Lambda_2 = \{(0, 1), (1, 2), (u, 2)\}$ forms a 3 edge. Therefore, there is no unit other than 1. Hence $F_{p^\alpha} = \mathbb{Z}_2$. Similar argument can be made for the case when $R_2 = \mathbb{Z}[x]/ < x^2 >$.

If both R_1 and R_2 are fields, then R is isomorphic to $F_{p^\alpha} \times F_{p^\beta}$ for some primes p, q. This completes the proof. □

We summarize Propositions 5, 7 and 8 in the following theorem:

Theorem 6 *Suppose R is a finite ring with $HG(R) = 2$. Then R is isomorphic with one of the following rings:*

$\mathbb{Z}_8, \mathbb{Z}_4[x]/ < x^2 + 2, 2x >, \mathbb{Z}_2[x]/ < x^3 >$.

$F_{p^\alpha}[x_1, x_2, \ldots, x_r]/ < x_i \cdot x_j >$ *where,* $1 \leq i, j \leq r$, *p is a prime,* $r \geq 1$ *and* F_{p^α} *is a finite field other then* $\mathbb{Z}_2$.

$\mathbb{Z}_2[x_1, x_2, \ldots, x_r]/ < x_i \cdot x_j >$ *where,* $1 \leq i, j \leq r$ *and* $r \geq 2$.

$\mathbb{Z}_{p^2}[\xi][x_1, x_2, \ldots, x_r]/ < x_i \cdot x_j,\ p \cdot x_j >$ *where,* $r \geq 0,\ 1 \leq i, j \leq r$ *and* $\mathbb{Z}_{p^2}[\xi]$ *is a coefficient ring in* R *other than* $\mathbb{Z}_4$.

$\mathbb{Z}_4[x_1, x_2, \ldots, x_r]/ < x_i \cdot x_j,\ 2 \cdot x_i >$ *where,* $r \geq 1,\ 1 \leq i, j \leq r$.

$F_{p^\alpha} \times F_{q^\beta}$, $\mathbb{Z}_2 \times \mathbb{Z}_4$ *or* $\mathbb{Z}_2 \times \mathbb{Z}_2[x]/ < x^2 >$ *where* $p,\ q$ *are primes.*

Acknowledgments Author wishes to thank referee for thorough reading of this paper and making numerous constructive recommendations, which has brought more clarity to this paper.

References

1. D.F. Anderson, P.S. Livingston, The zero-divisor graph of a commutative ring. J. Algebra **217**, 434–447 (1999)
2. D.D. Anderson, M. Naseer, Beck's coloring of a commutative ring. J. Algebra **159**, 500–514 (1993)
3. M.F. Atiyah, I.G. MacDonald, *Introduction to Commutative Algebra* (Addison-Wesley, Reading, 1969)
4. I. Beck, Coloring of commutative rings. J. Algebra **116**, 208–226 (1988)
5. C. Berge, *Hypergraphs: Combinatorics of Finite Sets* (North-Holland, Reading, 1989)
6. G. Bini, F. Flamini, *Finite Commutative Rings and Their Applications* (Kuwer Academic, Cambridge, 2002)
7. J. Coykendall, S. Sather-Wagstaff, L. Sheppardson, S. Spiroff, On zero divisor graphs. Prog. Commut. Algebra **2**, 241–299 (2012)
8. D. Eisenbud, *Commutative Algebra: With a View Toward Algebraic Geometry*. Graduate Texts in Mathematics, vol. 150 (Springer, Berlin, 1995)
9. C. Eslahchi, A. M. Rahimi, The k-zero-divisor hypergraph of a commutative ring. Int. J. Math. Math. Sci. **2007**, Art. ID 50875, 15 pp. MR2306364 (2007)
10. N. Ganesan, Properties of rings with a finite number of zero divisors-II. Math. Ann. **161**(4), 241–246 (1965)
11. J. Pathak, $k-$zero divisor hyper-graphs on commutative rings. J. Combin. Math. Combin. Comput. **102**, 99–108 (2017)
12. J. Pathak, The number of minimal prime ideals in a k-integral domain and an algorithm for constructing maximal k-zero divisors in Z_n. J. Combin. Math. Combin. Comput. **110**, 39–60 (2019)
13. J. Pathak, Commutative rings with finite hyper-graph and automorphisms of maximal hypergraphs on $\mathbb{Z}_n$. Math. Today **35**, 34–43 (2019)

been proposed. Despite this, modularity continues to be used and there is continued value in improving our understanding of the metric.

Many graphs have degree distributions with a heavy tail. Whether these arise from a power-law [2] or some other distribution [3], these graphs have some vertices with very large degree and many vertices with a low degree. Modularity explicitly treats vertices of different degrees differently, imposing a larger penalty when high-degree vertices are put in the same cluster. We posit that clustering algorithms that use modularity as the score will perform poorly on lower-degree vertices.

We propose a natural variant of modularity, *flat modularity*, which does not differentiate between the vertices based on their degrees. We use the Greedy Modularity algorithm [4] to cluster using modularity and also with our proposed modularity variant on LFR graphs that have degree distributions satisfying a power law. We compare the resulting clusters to the planted clusters and evaluate the results using the Matthews correlation coefficient. We observe that climbing using our variant score improves the match to the planted clusters. We also observe that the gain in performance seems to come by improving the clustering of low-degree vertices, see Fig. 3 on page 206.

2 Modularity

Given a division of the vertices of a graph into clusters, modularity compares the fraction of edges internal to clusters to the average value over all graphs with the same degree distribution (where we allow multiple edges and loops). One of the many ways to write the modularity of a given division into clusters is

$$Q = \frac{1}{2L} \sum_{v} \sum_{w} C_{vw} \left(A_{vw} - \frac{k_v k_w}{2L} \right) \tag{1}$$

where Q is the modularity, L the number of edges in the graph, k_v the degree of the vertex v, and A and C are adjacency matrices, A for the graph and C for the clusters; that is $C_{vw} = 1$ when v and w are in the same cluster and $C_{vw} = 0$ otherwise.

We interpret modularity as containing a bonus A_{vw} for putting adjacent vertices into the same cluster. The penalty $k_v k_w/(2L)$ encourages keeping vertices separate. This penalty therefore varies among the pairs of vertices and our primary question is whether or not this is desirable.

Modularity is the main ingredient in a number of algorithms intended to divide the vertices of a graph into clusters. We will use the Greedy Modularity algorithm [4] though the Louvain algorithm [5] is known to have a better run-time. A more recent example of such an algorithm can be found in [6]. Algorithms based on modularity continue to be used to identify clusterings [7]. We chose Greedy Modularity because it is a simple climb and we believe that this places more weight on the score being used than a more complex climb.

It is now known [8] that modularity has a fundamental drawback, called the *resolution limit*. Modularity tends to want to collapse clusters that are smaller than this resolution limit. A common workaround is to add a *resolution* r to the modularity, forming

$$Q_r = \frac{1}{2L} \sum_v \sum_w C_{vw} \left(r \cdot A_{vw} - \frac{k_v k_w}{2L} \right) \tag{2}$$

for a value r between 0 and 1; we reduce the reward for putting adjacent vertices into the same cluster. This increases the relative effect of the penalty and hence discourages combining communities.

A large number of other measures of the quality of a division into clusters have been proposed. Many are considered in [9] which also suggests different ways to evaluate these measures. We are going to use artifical graphs with planted clusters and that simplifies the evaluation phase.

3 Some Power-Law Graphs

The penalty term in the formula for modularity, (2), explicitly depends on the degrees of the vertices. In many real world graphs, for example those arising from power grids or cell interconnections, geometry forces all vertices to have degrees in a relatively small range. In this case, having the penalty depend on the degree will have only a minor effect. In other real world graphs, for example those arising from web page links or social networks, the degree distribution can be quite heavy-tailed (see [2] for further discussion). In this case, the magnitude of the penalty can cover a large range.

To study the consequences of the widely varying penalty, we will consider LFR graphs [10], a standard benchmark with planted clusters where both the degree distribution and cluster sizes follow a power law. Specifically, we use the implementation of LFR graphs in the NetworkX [11] function `LFR_benchmark_graph`. We will use `gamma` (that is, γ) for the exponent in the power law of the degree distribution and vary a random seed `seed` as we construct the graph. All of the other parameters are specified in Fig. 1. While one of our inputs seems to say that the maximum degree will be 50, in fact our graphs commonly have higher degrees. For one example we looked at, the degrees ran from 14 to 67. The input parameter with value 1000 represents the number of vertices in the graph. The input parameter with value 2 is the exponent in the power law for the community sizes. The parameter with value 0.5 is the mixing parameter; for each vertex roughly half of the edges will be within the planted cluster and half will be external. We believe that this setting means that recovering the planted clusters for the resulting graph is difficult.

```
LFR_benchmark_graph(1000, gamma, 2, 0.5, average_degree=20,
                    max_degree=50, min_community=20,
                    max_community=100, seed=seed)
```

Fig. 1 The command we use to generate our LFR benchmark graphs with inputs gamma and seed

For γ we use 2.5, 3.0, and 3.5. For each, we generate 1001 graphs using the seeds 0 through 1000. Later, we will also raise the mixing parameter to 0.6 to consider even harder problems where 60% of the edges are external to the clusters.

4 Finding Clusterings and Measuring Their Quality

Given an LFR graph, we wish to find a division of the vertices into clusters. We use the Greedy Modularity algorithm [4] to do so. This algorithm is fairly simple: start with the clustering of all vertices in their own cluster and, at each step, if combining any two clusters improves the modularity, combine the two that give the greatest modularity. Ties can arise; our implementation relies on sorting the improvements and so should return the same answer on each run—we suspect that this performs roughly the same as randomly breaking ties.

We chose this algorithm for its simplicity. Our intention is not to measure the capability of the algorithm but rather the score on which it is climbing. We hope that using a simple algorithm will emphasize the differences in the quality of the score. We assume that more involved algorithms, for example the Louvain algorithm [5], will also perform better given a better score.

Along with the authors of [12], we believe that the best way to evaluate our found clusters is to look at whether individual pairs are correctly clustered together or apart. To be explicit, each pair of vertices has a true label, clustered together or not, from the planted clusters as well as a learned label from the clusters we produced via Greedy Modularity. The $\binom{1000}{2}$ pairs of vertices are mostly not in the same cluster, so we have unbalanced class labels. Following the recommendation in [13] and elsewhere, we use the Matthews correlation coefficient [14] to turn the resulting 2×2 confusion matrix into a single number. We rely on the implementation of the Matthews correlation coefficient in Scikit-learn [15].

The Matthews correlation coefficient ranges from $[-1; 1]$ with 0 meant to represent a random distribution of the learned labels. Our learned clusters are better than random and so all of the MCC's we will see will be greater than 0.

5 Clustering Using Modularity

We begin by identifying the right resolution to use for our LFR graphs, Fig. 2 on the facing page. We have plotted the median and quartiles of the Matthews correlation

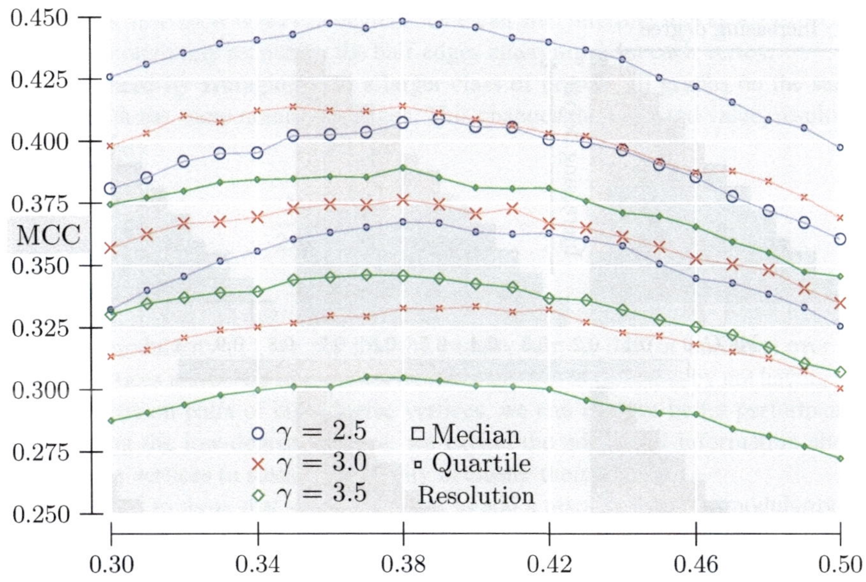

Fig. 2 The Matthews correlation coefficient for the clustering found by the Greedy Modularity algorithm using standard modularity Q_r (2) with resolution r. We present the median and quartile values of the MCC over 1001 LFR graphs with 1000 vertices and degree distributions with exponent γ; larger MCC is better

coefficient between the found clustering and the ground truth clustering; because we used 1001 graphs, these represent the results from particular graphs. Graphs with larger γ prove to be harder to cluster; we contend that this is because they contain a larger proportion of low degree vertices. We use the greatest median to identify the best choice of resolution, which is 0.39 for $\gamma = 2.5$, 0.38 for $\gamma = 3.0$, and 0.37 for $\gamma = 3.5$.

We looked further at the three graphs that represent the median and quartile scores for $r = 0.39$ with $\gamma = 2.5$. In the top row of Fig. 3 on the next page we plot the Matthews correlation coefficient between the true clusters and the recovered clusters when we restrict to pairs of vertices with certain degrees. For the leftmost figure, the bottom row represents vertices of degrees 14 through 18, of which there are 85. The lower left square represents all pairs of vertices with those degrees. The final box as we slide to the right in that triangle represents pairs where one vertex has degree 14 through 18 and the other degree 48 through 61; there are only 61 such vertices, so the block is narrower. Degrees are divided from smallest to largest with a break when the size of the current block would exceed 100 if we included the next degree. The next row up represents pairs involving vertices with degrees 19 or 20.

Looking at these results, a reasonable interpretation is that when both of the vertices have a relatively high degree, we can do pretty well at determining whether the two vertices should be clustered together or not. However, when at least one of

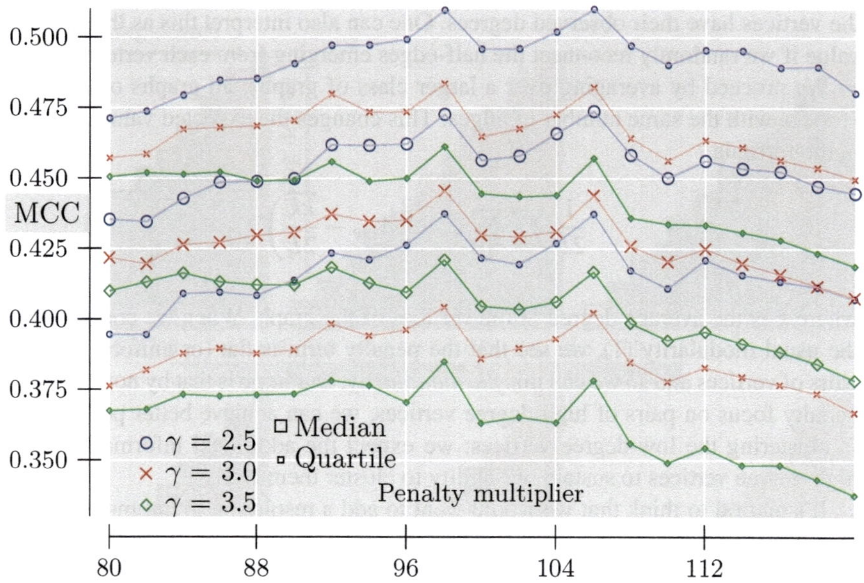

Fig. 4 The Matthews correlation coefficient for the clustering found by the Greedy Modularity algorithm using flat modularity Q_R^b (3) with penalty multiplier R. We present the median and quartile values of the MCC over 1001 LFR graphs with 1000 vertices and degree distributions with exponent γ; larger MCC is better

For the resolution, r, we considered values to two decimal places between 0 and 1. When we found that the optimal value of R would be around 100, we decided to restrict R to even integers between 0 and 200. The median and quartiles of the Matthews correlation coefficient when we apply the Greedy Modularity algorithm with flat modularity as the score are depicted in Fig. 4 on the following page.

Among other features, it is still the case that increasing γ hurts our ability to recover the true clustering. The scores are also less smooth as a function of the penalty multiplier. This might be because there are more ties now; for modularity, the varying penalty is likely to drive down the number of ties. We collect the best scores we found in Table 1 on the next page. Using flat modularity has improved the fidelity to the planted clustering and the relative decrease in performance with rising γ is less when we use flat modularity.

The second row of Fig. 3 on page 206 depicts the results of clustering using flat modularity on the same three graphs whose results in clustering using modularity are depicted in the first row. Switching to flat modularity does not hurt the performance in clustering high-degree vertices but does improve clustering of low-degree vertices. This is particularly evident in the lower right of the triangles which have become much darker. Therefore, we have done a better job matching low-degree vertices with high-degree vertices. The lower lefts of the triangles indicate

Table 1 Median and quartiles of the Matthews correlation coefficient using modularity with the best resolution we found and using flat modularity with the best penalty multiplier we found; a larger MCC is better

	Modularity				Flat modularity			
γ	r	1/4-ile	Median	3/4-ile	R	1/4-ile	Median	3/4-ile
2.5	0.39	0.3670	0.4091	0.4470	98	0.4373	0.4727	0.5096
3.0	0.38	0.3327	0.3762	0.4140	98	0.4042	0.4455	0.4834
3.5	0.37	0.3052	0.3462	0.3856	98	0.3849	0.4210	0.4612

Table 2 Median and quartiles of the Matthews correlation coefficient over pairs where one vertex has degree at most 20 and one degree at least 40 split between climbing using modularity with the best resolution we found and using flat modularity with the best penalty multiplier we found; a larger MCC is better

	Modularity				Flat modularity			
γ	r	1/4-ile	Median	3/4-ile	R	1/4-ile	Median	3/4-ile
2.5	0.39	0.1705	0.2115	0.2538	98	0.2243	0.2661	0.3142
3.0	0.38	0.1774	0.2154	0.2607	98	0.2283	0.2702	0.3189
3.5	0.37	0.1798	0.2224	0.2647	98	0.2266	0.2701	0.3216

there is still room for improvement in properly matching low-degree vertices amongst themselves, though we note that this is likely to be difficult in general.

These observations generally hold for all of the graphs. We computed the Matthews correlation coefficient between the found clustering and the true clustering restricted to pairs of vertices where one has degree at most 20 and one degree at least 40. This corresponds to the lower right of our triangles. The medians and quartiles over the 1001 graphs are found in Table 2; the r and R values are the ones from Table 1, we did not attempt to find the best r and R for this set of vertices. This supports the claim that using flat modularity leads to improved clustering of low-degree vertices. Note that the observed values for these pairs are more similar for the different values of γ than what we observed in Table 1. We posit that there is relatively consistent performance among pairs of high-degree vertices, among pairs of low-degree vertices, and among pairs where one is low-degree and the other high-degree. Therefore, the change in performance with different γ's is due to the change in the numbers of each type of pair.

For each of the 1001 graphs with $\gamma = 2.5$, we have plotted the Matthews correlation coefficient of the found clustering using standard modularity with resolution $r = 0.39$ against the Matthews correlation coefficient from using flat modularity with multiplier $R = 98$ in Fig. 5 on the following page. We do not see, for example, that the improvement only comes from graphs where the original climb did particularly poorly; rather there is general improvement. We have not done any analysis to identify features of the graphs that might correlate with either score or the difference in the scores.

We repeated these experiments with the mixing parameter set to 0.6 instead of the 0.5 of the main text. We record the results that correspond to Tables 1 and 2 in

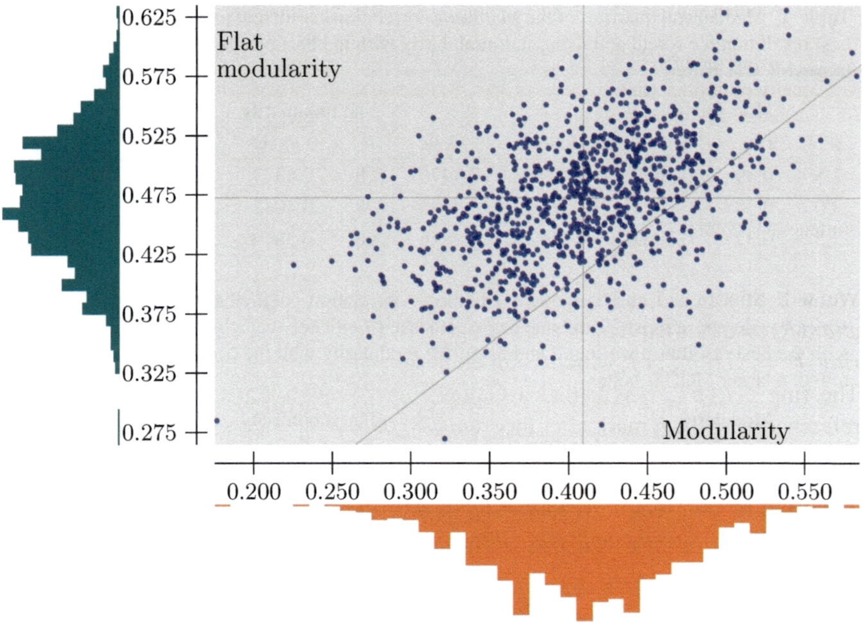

Fig. 5 For the 1001 LFR graphs we considered with $\gamma = 2.5$, we plot the Matthews correlation coefficient for the clustering found using modularity (x-axis; Q_r of (2) with $r = 0.39$) against the Matthews correlation coefficient for the clustering found using flat modularity (y-axis; $Q^{\flat}_R$ of (3) with $R = 98$). A larger MCC is better. The faint horizontal and vertical lines represent the medians of the marginal distributions and the faint diagonal line is $y = x$

Table 3 on page 211. As in the 0.5 case, we chose r and R to maximize the median Matthews correlation coefficient between the planted clusters and the found clusters over all of the vertices and, for those values of r and R, we report out statistics for the Matthews correlation coefficient between the planted clustering and the found clustering restricted to pairs of vertices where one vertex has degree at most 20 and one degree at least 40. In this harder case, there is again a large difference in favor of flat modularity in both the overall scores and restricted to pairs that involve one low-degree vertex and one high-degree vertex.

7 Conclusions

Modularity continues to be a factor in finding clusters in graphs. If we interpret modularity as involving a bonus and a penalty, then the penalty is higher for higher-degree vertices. When we use a simple clustering algorithm that attempts to maximize modularity, it fares far better on the high-degree vertices than on the low-degree vertices.

Table 3 Median and quartiles of the Matthews correlation coefficient for either all pairs of vertices or only pairs where one vertex has degree at most 20 and one degree at least 40 split between climbing using modularity with the best resolution we found and using flat modularity with the best penalty multiplier we found over 1001 graphs where the mixing parameter was 0.6; a larger MCC is better. The chosen values of r and R maximize the median Matthews correlation over all pairs of vertices

	Modularity				Flat modularity			
γ	r	1/4-ile	Median	3/4-ile	R	1/4-ile	Median	3/4-ile
All pairs								
2.5	0.39	0.0792	0.1072	0.1322	106	0.1535	0.1880	0.2186
3.0	0.39	0.0723	0.0949	0.1177	106	0.1386	0.1700	0.2032
3.5	0.44	0.0630	0.0861	0.1060	106	0.1225	0.1529	0.1826
Pairs with a low-degree vertex and a high-degree vertex only								
2.5	0.39	0.0188	0.0271	0.0383	106	0.0328	0.0468	0.0649
3.0	0.39	0.0231	0.0334	0.0465	106	0.0379	0.0535	0.0728
3.5	0.44	0.0263	0.0363	0.0484	106	0.0422	0.0588	0.0804

We therefore ask whether it is better to have the penalty be independent of the degree of the vertex. We found that the same simple algorithm, now using flat modularity, finds clusterings that better match the ground truth, in particular improving the clustering of lower-degree vertices. Meanwhile, our implementation became simpler because we no longer needed to track the degrees of the vertices.

We have more information about the vertices with higher degrees and so it is not surprising that we can cluster them more effectively. Having our penalty focus on this high information region proves less helpful than flattening the penalty. Perhaps one could even go further and try to apply the penalty disproportionately to low-degree vertices; we have not tested this idea. Our choice of penaly corresponds to a different random model; it would be of interest to find a random model that would apply the penaly disproportionately to low-degree vertices.

While our tests were confined to graphs in which the vertex degrees satisfy a power law, we expect the results to generalize to other heavy-tailed degree distributions. Since such graphs occur frequently, and finding communities in graphs continues to be an important problem, we believe that the simple change from modularity to flat modularity can improve a broad range of clustering results with minimal change to implementations.

References

1. M.E.J. Newman, M. Girvan, Finding and evaluating community structure in networks. Phys. Rev. E. **69**, 026113 (2004)
2. A.-L. Barabási, *Network Science* (Cambridge University Press, Cambridge, 2016)
3. A. Clauset, C.R. Shalizi, M.E.J. Newman, Power-law distributions in empirical data. SIAM Rev. **51**, 661–703 (2009)

GPSR Compliance

The European Union's (EU) General Product Safety Regulation (GPSR) is a set of rules that requires consumer products to be safe and our obligations to ensure this.

If you have any concerns about our products, you can contact us on ProductSafety@springernature.com

In case Publisher is established outside the EU, the EU authorized representative is:

Springer Nature Customer Service Center GmbH
Europaplatz 3
69115 Heidelberg, Germany

Batch number: 10732090

Printed by Printforce, the Netherlands